PRAISE FOR *THREE SCIEN*...
BY ROBERT...

"Makes entropy as absorbing as a gossip column . . . Mr. Wright is a master of the intellectual profile."
—*Wall Street Journal*

"A gold mine and a mine field of ideas . . . *Three Scientists and Their Gods* is a book designed to rattle people, especially people who like to think."
—*San Francisco Chronicle*

"Wright, in a fine prose style equal to the demands of elucidating complex science, builds an exciting narrative around extended profiles of three fascinating scientists."
—*Cleveland Plain Dealer*

"Robert Wright has written an exceptionally thoughtful and conceptually brave book that tackles not only some complex contentions of science and religion but also basic notions of philosophy and logic. The result is wonderfully and eminently readable. . . . Bravo for this first step."
—*Washington Monthly*

"This is science—and science writing—with a refreshingly human face."
—*Wilson Quarterly*

"This is a wonderful, thought-provoking book."
—*Publishers Weekly*

"A surprisingly deep and witty book."
—*Kirkus*

"If you've never read a whole science book at one sitting, *Three Scientists* may change that."
—*Philadelphia Inquirer*

/

T H R E E
SCIENTISTS
A N D   T H E I R
G O D S

LOOKING FOR MEANING
IN AN AGE OF INFORMATION

ROBERT WRIGHT

PERENNIAL LIBRARY

Harper & Row, Publishers, New York
Grand Rapids, Philadelphia, St. Louis, San Francisco
London, Singapore, Sydney, Tokyo

First PERENNIAL LIBRARY edition published 1989

Library of Congress Cataloging-in-Publication Data
Wright, Robert, 1957–
    Three scientists and their gods : looking for meaning in an age of information / Robert Wright.
        p.   cm.
    Bibliography: p.
    Includes index.
    ISBN 0-06-097257-2
    1. Information theory.   I. Title.
[Q360.W75   1989]
003'.54—dc20
                                                                89-45119
                                                                CIP

89 90 91 92 93 FG 10 9 8 7 6 5 4 3 2 1

For my parents,
Raymond Jay Wright
(February 14, 1919, *to* September 14, 1987)
and
Margie Moselle Wright

ACKNOWLEDGMENTS

It is literally impossible to thank all the people who have contributed to the writing of this book, so I will focus on those whose contributions were fairly direct. First, I am grateful to the three stars of the show—Ed Fredkin, Edward O. Wilson, and Kenneth Boulding—for submitting to hours and hours of observation and interrogation in various locales. In addition, all agreed to read long portions of early drafts of the profiles with no assurance that I would make any changes they recommended.

I am indebted also to a number of other people who read and critiqued one or more chapters in draft form: Richard D. Alexander, Charles H. Bennett, John Tyler Bonner, Richard Dawkins, Rolf Landauer, Norman Margolus, Philip Hefner, Alan Peterkofsky, and Thomas Sebeok. And James R. Beniger deserves special thanks. In addition to ruthlessly marking up two chapters, he has spent a number of hours arguing (and even, at times, agreeing) with me about ideas central to this book.

I want also to thank Paul Libassi, editor of *The Sciences*, for letting, and helping, me write the column that led indirectly to this book; Hugh O'Neill, of Times Books, for taking a gamble on me; Joel Nanni and Mark Graven for years ago steering me toward some of the terrain this book covers; Eva-Lynne Greene for tirelessly tracking down obscure journal articles and checking suspect facts; Ullica Segerstrale for sending me her dissertation on sociobiology, which, I understand, she is adapting for publication by the University of Chicago Press; Horton Johnson for several stimulating discussions; and Randall Rothenberg for early advice. I should also thank someone I've never communicated with: Cynthia Earl Kerman, a former secretary of Kenneth Boulding's whose doctoral dissertation on him became the biography *Creative Tension*, published by the University of Michigan Press. This book, with its exhaustive factual basis, was an invaluable reference source, and Kerman's interpretation of Boulding's basic intellectual character was quite enlightening.

A number of friends helped sustain me during the ordeal of writing a book, and, as a bonus, many of them responded repeatedly to the question,

"What do you think of *this* title?" They include Don Arbour, Sarah Boxer, Jane Epstein, Hank Hersch, Susan Korones, Gordie Kotik, Steve Nelson, Alex Wolff, and, above all, Lisa O'Neill. Finally, I owe a large debt to Gary Krist, who slogged through the entire manuscript without complaining, improved it considerably with his reactions, and helped keep my faith in the project intact.

Of course, notwithstanding the influence of all these people, I alone am responsible for any factual errors or other flaws in this book.

CONTENTS

A NOTE TO READERS

I don't want to alarm you, but this book is about—

1. the concept of information;

2. the concepts of meaning and purpose, in both their mundane and cosmic senses;

3. the function of information at various levels of organic organization (in bacteria, ant colonies, human brains, and supermarket chains, for example), with particular emphasis on its role in reconciling life with the second law of thermodynamics;

4. the meaning of the information age, viewed in light of the role information has played throughout evolution;

5. the meaning of life; and

6. a couple of other issues at the intersection of religion and science.

Now for the good news: this book is also about three living, breathing, and, I think, unusually interesting human beings. In fact, they are what the book is *mainly* about. So, for the most part, all you have to do is read about them—about their personal histories, their ways of living, and their very ambitious ways of thinking about the universe and our place in it—and let the above subjects emerge in the process. It will be fairly painless, as these things go.

—R. W.

PART I

EDWARD

FREDKIN

CHAPTER
ONE

FLYING SOLO

Ed Fredkin scans the visual field systematically. He checks the instrument panel regularly. He makes conversation sparingly. He is cool, collected, in control. He is the optimally efficient pilot.

The plane is a Cessna Stationair Six—a six-passenger, single-engine, amphibious plane, the kind with the wheels recessed in pontoons. Fredkin bought it not long ago, for $165,000, and is still working out a few kinks; right now he is taking it for a spin around the neighborhood in the wake of minor mechanical work.

He points down at several brown-green masses of land embedded in a turquoise sea so clear that the shadows of yachts are distinctly visible on its sandy bottom. He singles out a small island with a good-sized villa and a swimming pool. That, he explains, cost $4.5 million to build. The compound, and the island as well, belong to "the guy that owns Boy George"—the rock star's agent, or manager, or something.

I remark, loudly enough to overcome the engine noise, "It's nice."

Yes, Ed allows, it's nice. He adds, "It's not as nice as my island."

He's joking, I guess, but it turns out he's right. Ed Fredkin's island, which will come into view momentarily, is bigger and prettier. It is about 125 acres, and the hill that constitutes its bulk is a deep green—a mixture of reeds and cacti, sea grape and turpentine trees, machineel and frangipani. Its several beaches—pale lines from up here—range from prosaic to sublime, and the coral within their shallow waters attracts little and big fish whose colors look as if they were coordinated by Alexander Julian. On the island's west side are immense rocks, suitable for careful climbing, and on the east side is a bar and restaurant and a modest hotel: three clapboard buildings, each with a few

3

rooms. Between east and west is the secluded island villa, where Ed
and his family are staying this week. It is a handsome structure, even
if, with a construction cost of half a million dollars, it is not in the
same league as the villa of the guy that owns Boy George. All told,
Moskito Island—or Drake's Anchorage, as the brochures call it—is a
nice place for Fredkin to spend the few weeks each year when he is
not up in the Boston area tending his various other businesses. And
it is a remarkable asset for a man who was born into a dirt-poor,
Depression-era family to have acquired.

In addition to being a self-made millionaire, Fredkin is a self-made
intellectual. About twenty years ago, at age thirty-four, without so
much as a bachelor's degree to his name, he became a full professor
at the Massachusetts Institute of Technology. And he did not rest on
this laurel. Though hired to teach computer science (and then selected
to guide MIT's now-eminent computer science laboratory through
some of its formative years), he soon branched out into more offbeat
things. Perhaps the most idiosyncratic of the several idiosyncratic
courses he originated was one on physics—"digital physics"—in
which he propounded the most idiosyncratic of his several idiosyncratic theories. This theory is the reason I'm here, hovering over the
British Virgin Islands.

Fredkin's theory is one of those things you have to prepare people
for. You have to say, "Now this is going to sound pretty weird, and
in a way it is, but in a way it's not as weird as it sounds, and you'll
see this once you understand it, but that may take a while, so in the
meantime, don't prejudge it, and don't casually dismiss it."

Ed Fredkin thinks the universe is a computer. A really big
one.

Fredkin works in a twilight zone of modern science—the interface
between computer science and physics. Here the two concepts that
traditionally have ranked among science's most fundamental—matter
and energy—keep bumping into a third: information. The exact
relationship among the three is a question whose implications can,
depending on how you define *information*, stretch far beyond physics
and computer science, into biology and the social sciences. And, to
date, it is a question without a clear answer, a question vague
enough, and basic enough, to have inspired a wide variety of opinions.

Some scientists have settled for modest and sober answers. Information, they will tell you, is just one of many derivatives of matter and energy; it is embodied in things such as a computer's electrons and a brain's neural firings, things such as newsprint and radio waves; and that is that. Others talk in grander terms, suggesting that information deserves full equality with matter and energy, that it should join them in some sort of scientific trinity, that these three things are the main ingredients of reality.

Fredkin goes further still. According to his theory of digital physics, information is *more* fundamental than matter and energy. He believes that atoms, electrons, and quarks consist ultimately of bits—binary units of information, like those that are the currency of computation in a personal computer or a pocket calculator. And he believes that the behavior of those bits, and thus of the entire universe, is governed by a single programming rule—"the cause and prime mover of everything," he calls it.

Among the scientists who don't dismiss Fredkin out of hand are some very smart people. Marvin Minsky, a computer scientist (and polymath) at MIT whose renown approaches cultic proportions in some circles, calls Fredkin "Einstein-like" in his ability to find deep principles through simple intellectual excursions. If it is true that most physicists think Fredkin is off the wall, said Minsky, it is also true that "most physicists are the ones who don't invent new theories"; they go about their work with tunnel vision, never questioning the dogma of the day. When it comes to the kind of basic reformulation of thought proposed by Fredkin, Minsky said, "there's no point in talking to anyone but a Feynman or an Einstein or a Pauli. The rest are just Republicans and Democrats."

Richard Feynman, the late Nobel laureate who taught at the California Institute of Technology, considered Fredkin a brilliant and consistently original, though sometimes incautious, thinker. If anyone is going to come up with a new and fruitful way of looking at physics, Feynman told me, Fredkin will.

Notwithstanding their moral support, though, neither Feynman nor Minsky was ever convinced that the universe is a computer. They were endorsing Fredkin's mind, not this particular manifestation of it. When it comes to digital physics, Ed Fredkin is flying solo.

He knows this, and he regrets that his ideas continue to lack the respect of his peers. But his self-confidence is unshaken. You see, Fredkin explains, he has had an odd childhood, and an odd education, and an odd career, all of which have endowed him with an odd perspective, from which the essential nature of the universe happens to be clearly visible. "I feel like I'm the only person with eyes in a world that's blind," he says.

CHAPTER

TWO

FREDKIN'S ISLAND

Mealtime at the restaurant on Ed Fredkin's island gives you some idea of what wealth would be like. About half the dishes on the menu go by foreign names, and even the English entrées have an exotic air. Witness Dolphin in Curry Sauce with Bananas. The food is prepared by a large man named Brutus and is humbly submitted to diners by men and women native to nearby islands. The restaurant is open-air, ventilated by a sea breeze that is warm during the day, cool at night, and almost always moist. Between diners and the ocean is a knee-high stone wall, against which waves lap rhythmically. Beyond are other islands and a horizon typically blanketed by cottony clouds. Above is a thatched ceiling, concealing, if the truth be told, a sheet of corrugated steel, which rests on thick vertical wooden posts veneered with cross sections of large rocks. Propped against a post here and there are impressive relics—an old, rusting anchor, a piece of coral that bears an uncanny resemblance to a gargantuan mushroom. Hanging overhead is a long, weathered plank with H.E. THOMPSO impressed upon it—90 percent of a sunken sailing ship's nameplate.

It is lunchtime now, and Ed is sitting in a cane-and-wicker chair across the table from me, wearing a light cotton sport shirt and gray swimming trunks. He was out trying to windsurf this morning, and he enjoyed only the marginal success that one would predict on the basis of his appearance. He is fairly tall and very thin, and has a softness about him—not effeminacy, but a gentleness of expression and manner—and the complexion of a scholar; even after a week on the island, his face doesn't vary much from white, except for his nose, which turns red. The plastic frames of his glasses, in a modified aviator configuration, surround narrow eyes; there are times—early in the morning or right after a nap—when they barely qualify as slits. His hair, perennially semi-combed, is black with a little gray.

Ed is a pleasant mealtime companion. He has much to say that is interesting, and he is good at talking. This is fortunate, because generally he does most of it; he has little curiosity about other people's minds, unless their interests happen to coincide with his, which few people's do. "He's right above us," his wife, Joyce, once explained to me, holding her left hand just above her head, parallel to the ground. "Right here looking down. He's not looking down saying, 'I know more than you.' He's just going along his own way."

Joyce is sitting to Ed's left. As is often the case, she is wearing something made of white cotton—a jumpsuit today. It nicely sets off her deep black skin. Joyce is very attractive, and her beauty is enhanced by her accent, which sounds somehow more refined, more aristocratic, than that of the waiters and waitresses. It is misleading. She grew up in a poor neighborhood on the island of St. Thomas in the American Virgin Islands, daughter of a bartender whom she describes as alcoholic and mercurial. He was hardworking, though, and laid the foundation for a better life, sending Joyce to private school and then to a small college near Boston.

If Joyce has made the socioeconomic leap with anything less than complete grace, it is not for lack of effort. She uses the word *exquisite* several times a day, usually in reference to food, and she buys clothes in France. When entertaining guests (last night, the lieutenant governor of the British Virgin Islands was among them), she observes protocol painstakingly. She solemnly sips the inaugural glass of champagne, cocks her head slightly, contemplating its quality, and then nods her approval to the waiter.

When Joyce met Ed, in 1977, she was twenty years old, not quite half his age. He was still in his 1960s mode then, with hair that nearly reached his shoulders. Joyce's sister characterized him as a "big-nosed, long-haired hippie." Joyce was forced to concede the point, but its truth did not lessen her ardor. "I was just into his brain," she recalls.

To Joyce's left, at the end of the table, is Ed's sister, Joan, who is making her first visit to the island in about a decade. She is pretty: dark hair, rich, dark eyes, a fine nose, and a look of health about her. Joan is Ed's senior by only sixteen months, and among the attendant tensions between them is a philosophical one: she thinks that science, with its reliance on "sequential" thought, is missing something important. She believes that *some* odd coincidences are no accident and that, more generally, there are a lot of weird things going on in the world.

To Joan's left, across the table from Joyce, is John Macone, an accomplished pilot of airplanes and balloons, and the man who taught Ed to fly a seaplane. Macone now heads a small, Fredkin-financed company that specializes in "reverse-osmosis energy recovery"—or words to that effect.

I am sitting to Macone's left, and the seat to my left, at the end of the table, is empty. It soon will be occupied by a force to be reckoned with: little Richard, Ed's first child by this, his second, marriage. Richard is a milk-chocolate-brown child who is cute enough to become a media phenomenon. On the day of this lunch, he is three years old, going on four. He could pass for five on the basis of appearance, and six or seven on the basis of vocabulary. Just the other day, after Aunt Joan had spent a fair amount of time and affection on him, he looked up at her appreciatively and said, "I appreciate you." Richard would probably be precocious even if it weren't expected of him, but it emphatically is. "I suspect that if I ever do become famous," Ed has said, "it will be for being Richard's father."

Richard occupies an enviable position on Moskito Island. Drake's Anchorage is staffed by about twenty people—ten or so are on duty during the day—and they naturally are eager to please Joyce and Ed. Joyce and Ed—particularly Ed—are in turn eager to please Richard. And even if they weren't, Richard's debating skills would ensure that he prevailed in a fair number of disputes over resource allocation. The upshot is that, at age three, Richard Fredkin may be the youngest man ever to have secured political control of an inhabited island.

The food has not arrived, and Ed is passing time by explaining the value of looking at the world the way he does. "There's three great philosophical questions," he begins. "What is life? What is consciousness and thinking and memory and all that? And how does the universe work?" He says that his "informational viewpoint" encompasses all three. Take life, for example. Deoxyribonucleic acid, the material of heredity, is "a good example of digitally encoded information," he says. "The information that implies what a creature or a plant is going to be is encoded; it has its representation in the

DNA, right? Okay, now there is a process that takes that information and transforms it into the creature, okay?" His point is that a mouse, for example, is "a big, complicated informational process."

Ed exudes rationality. His voice isn't quite as even and precise as Mr. Spock's, but it's close, and the parallels don't end there. Ed rarely laughs or displays emotion—except, perhaps, the slightest sign of irritation under the most trying circumstances. He has never seen a problem that didn't have a perfectly logical solution, and he believes strongly that intelligence can be mechanized without limit. More than twenty years ago, he founded the Fredkin Prize, a $100,000 award for the creator of the first computer program to beat the world chess champion. He's thinking about raising it to a million.

Richard, who has spent most of the morning watching cartoons on a videotape player, walks up to the end of the table and posts himself between Ed and me. "I'm *very* hungry," he says.

Macone looks at Ed. "You've got to figure out how to transmit nourishment through a cathode ray tube. Then we'll have things all taken care of. Kids won't have to leave the TV set."

Ed is taking Richard a bit more seriously. "What would you like for lunch? Ham sandwich? Hot dog? Spaghetti?"

Richard thinks for a second. "French toast," he says, delighted with the selection.

Ed isn't sure they make French toast this late in the day. "Go in the kitchen and negotiate," he suggests.

"Say please," adds Joyce as Richard trots toward the kitchen.

Ed is not exactly the first person to have said that DNA consists of information, or that organic growth depends on intricate communication among cells, and I'm wondering why he has made the point sound so earthshaking. "That's conventionally accepted, right?" I ask.

"It wasn't when I started saying it." This is a recurring theme: when Fredkin's world view crystallized a quarter of a century ago, he immediately saw dozens of large-scale implications, ranging from physics to biology to psychology; a number of these ideas have gained currency since then, and he considers this trend an ongoing substantiation of his entire outlook.

Fredkin talks some more and then recaps. "What I'm saying is that, at the most basic level of complexity, an information process runs what we think of as physics. At the much higher level of complexity, life, DNA—you know, the biochemical functions—are controlled by a digital information process. Then, at another level, our thought processes are basically information processing." That is not to say, he stresses, that *everything* is *best* viewed as information. "It's just like there's mathematics and all these other things, but not everything is best viewed from a mathematical viewpoint. So what's being said is not that this comes along and replaces everything. It's one more avenue of modeling reality, and it happens to cover the sort of three biggest philosophical mysteries. So it sort of completes the picture."

A scream stops the conversation. It's from the kitchen, and it sounds like Richard's. He screams again, louder. One more time, the loudest yet. Now the sound of kitchen workers trying to soothe him. Joyce is dispatched to the scene.

"Probably hurt himself," I venture. "It sounded pretty sudden."

"No," Ed says with certainty. "He doesn't cry when he hurts himself. He's crying because his feelings are hurt. He probably asked for French toast and they probably told him they only serve it at breakfast. That's my guess."

"That's a major crisis," Macone observes.

Over the past two days, I have been compiling a mental list of staff members and island visitors who seem secretly annoyed by the amount of attention Richard receives. Macone's remark about the cathode ray tube was significant in this respect, and this last comment has secured him a spot on my roster.

Before Joyce makes it to the kitchen, Richard emerges and makes a beeline for the most sympathetic ear. "Daddy, the guys in the kitchen are doing my *co*coa wrong."

"They're doing your *co*coa wrong?" Ed asks in that voice parents use in such situations. "Well, did you explain to them how they're supposed to do it?"

"They don't know what I'm saying," says Richard, who, like many three-year-olds, does not always speak crisply.

"They won't listen to you?" Ed asks.

"*No.*"

One of the guilty parties is now standing next to Richard, awaiting orders. Ed gives them. "What he's trying to say is one teaspoon of

cocoa, three teaspoons of sugar in the bottom of a glass, and then—
he knows how to do it—you add a little drop of milk, stir it up, and
then add more milk." He looks down at Richard. "You want to show
them?"

Richard goes to show them, and Joyce calls after him, "One spoon
of cocoa, three of sugar."

"He knows that," Ed says sternly, as if personally insulted by her
doubt. "He knows how to do it. He makes a mess, but he makes
good cocoa." Ed reflects for a few seconds. "See, that was my great
trauma as a child," he says. "They kept treating me like a child. It
drove me crazy. They wouldn't listen to me."

Joan weighs in to the conversation. For what is no doubt not the
first time, she begs to differ with her brother. "Well, I think most
children suffered that," she suggests.

"Some suffered more than others," Ed replies. "Some *have* to suffer
more than others. I can prove that mathematically by Brouwer's fixed-
point theorem. Someone has to suffer most."

Invoking mathematics is not the way to win an argument with
Joan. She says, "Those who are not in power suffer that experience,
whether they're little, big, or whatever they are."

Ed pursues the point a ways and then turns to me. "See, my big
problem is being right. My sister, of course, could never agree with
that, but—" An old joke pops into his head. "I was wrong once," he
admits.

Macone plays straight man: "When?"

"What happened is, I once thought for a moment that I was wrong,
but I was mistaken."

After some polite laughter, the conversation starts to wander, but
Joan puts it back on track. "My father has that trait very well developed—being right."

"The difference is," says Ed, "he's not right as often."

"Oh," Joan says sarcastically.

Edward Fredkin's father, Manuel, was born in Chernigov, Russia,
in 1900. A decade later, the eldest of Manuel's brothers emigrated
to America, opened a combination grocery store and gas station,
stretched it into a chain, and in the process acquired land that hap-

pened to cover oil. He sent back word that America was hospitable toward industrious young men, and, some time after the Russian Revolution, Manuel followed. Manuel made his money on the frontiers of technology, selling radios—large, elaborate, cabinet radios. He opened one store, then another, then another. By the late 1920s, the M. S. Fredkin Company was thriving, and M. S. Fredkin—along with his wife, a concert pianist who had emigrated independently from Russia—was living the American Dream. Then came the Depression. The company slowly died, and Manuel began taking whatever work he could find. He typically held down two jobs at a time, putting in seventy or more hours a week.

Edward was born in 1934, the last of three children. He remembers his parents' arguing over money, and he remembers the series of modest houses they rented in East Hollywood's Los Feliz section. The Fredkins learned economy, and Ed has not forgotten it. He can reach into his pocket, pull out a Kleenex that should have been retired weeks before, and, with the help of cleaning solution, make an entire airplane windshield clear. He can take even a well-written computer program, sift through it for superfluous instructions, and edit accordingly, reducing both its size and its running time.

Manuel was by all accounts a competitive man, and he focused his competitive energy on the two boys: Edward and his older brother, Norman. "Oh, God, it was terrible," Joan remembers. "Even when Norman's shoes were getting to be the same size as his, he would not accept that." Norman's theory is that his father, though bright, was intellectually insecure; he seemed somehow threatened by the knowledge that the boys brought home from school. Ed exactly remembers his father's ritual gibe: "I have more brains in my little toe than you will ever have in your head."

Attempts to prove otherwise were doomed. One day, when Ed was about ten, the two of them were talking about the moon, and Ed decided to impress his father with a recently gleaned fact. The moon, he noted, is 240,000 miles from the earth. No, his father said, the figure is closer to 360,000. Ed, having just seen the number in an encyclopedia, was confident he was right, and he told his father as much. Manuel was not persuaded; 360,000 was the figure. Finally Ed brought out the encyclopedia and pointed to the number. His father slammed the book shut. "The encyclopedia's wrong," he said.

The issue lay dormant for a few days before being revived by Manuel. As Ed tells it: "He said, 'You remember we had this discussion'— and he always called it a discussion, not an argument— 'about how far the moon is from the earth?' And I thought, my God, he's going to admit he was wrong. And he says, 'Well, I was right. It's 360,000 versts from the earth.' I said, 'A *verst*? What's that?' And he said, 'That's a Russian mile.'" That was about as close as Manuel came to admitting error. His mistrust of books, experts, and all other sources of received wisdom was absorbed by Ed.

So was his competitiveness. "Ed, ever since I can remember, always could top any statement you could make," recalls Bill Fletcher, a childhood friend who was best man at Ed's first wedding. "If you said you just ran up three steps at a time, he had run up four steps at a time. If you just did fifteen push-ups, he had done sixteen."

Ed customarily considered himself the smartest kid in his class, and he used to place bets with friends on the outcome of tests. One such test, in fifth grade, proved pivotal. It consisted of ten questions, and one in particular he found worrisome. He had already noticed that the teacher misunderstood some concept, and this question embodied her misunderstanding; it was clear that what she wanted was the wrong answer. Ed deliberated for some time and finally decided to put down the genuinely correct answer. When the test came back a few days later, he found that the teacher had not stumbled onto the truth in the meanwhile; he had gotten a 90, which placed him in a dead heat with his fellow bettor. Ed now had no choice; he was morally and financially compelled to enlighten his teacher. She resisted at first, but finally relented, snatching the paper and promising to return it the next day. ("I was always amazed that when I would explain something like that to someone they weren't happy to know what the truth was.") Back the paper came, with the controversial problem marked correct but a "90" still sitting at the top. Asked for an explanation, the teacher pointed to the illegible scrawl in the upper left-hand corner; she had docked him ten points for misspelling his name. Ed was unruffled. All right, he said, he had misspelled his name—but if his name qualified as a test item, then there were eleven items, not ten, and he should get a 91. The beauty of his logic was lost on her.

This episode, coming after several years of dull and uninspiring teachers, was the last straw. "I said, 'Okay, so they want to play games instead of get at the truth? I'll play a good game.' So I started a new way of taking tests." All multiple-choice tests, whatever their nominal subject, became tests in psychology; success depended mainly on reading the intentions of test designers—realizing, for instance, that when forced, time and again, to fabricate four wrong answers for every right one, they will occasionally resort to making one wrong answer a mere paraphrasal of another; thus, if two of five answers are equal in meaning, both can be eliminated. By the time Ed went through cadet training in the air force, outsmarting test givers had grown into an exact science. "We had a multiple-choice exam, and my bet was that I could pass if I were only shown the answers and none of the questions. Some of the answers went, like, 347, 492, 513, 629—and I did pass the test."

So Ed learned to pass tests; he passed enough tests to get into and out of high school. But he never really learned how to study, and he amassed little academic evidence of ambition. That is not to say he had none. On the contrary, from a young age he had wanted to become a great scientist and to own the Empire State Building (which, he realized, represented a lot of rent). It was just that he had an aversion to some traditional prerequisites for success: treating figures of authority with respect, doing work to demonstrate to someone else that you know what you already know you know, etc.

Being a derelict student would have been easier had it not been for Norman. Norman was perfect.

The question has been put on the table (so has the food—mostly hamburgers, possibly the leanest I have ever eaten): How were Ed and Norman different as children? Ed defers to Joan, and Joan doesn't know where to begin, so vast were the differences. "Well, begin somewhere," Ed suggests forcefully.

Joan looks down and fiddles with her silverware. "He was, uh, more expressive, more sexually expressive, at an earlier age than you."

Ed gets to the heart of the matter: "You mean he dated more."

No, says Joan, it's more complicated than that. She looks at me. "I think that, uh, this part of Edward"—she points to her head—"developed at a young age."

Ed stands by his initial interpretation. "In other words, Norman was—what was the word? A stud or something . . . Not quite . . . It was close to that—very active in dating."

"Did he play sports?" I ask.

Yes, Ed says, Norman was on the gymnastics team.

Joan is still struggling to articulate her version of things. "He was more 'out there,' earlier—"

"More normal," Ed translates.

Norman was "the kid that would excel in school," Joan says. "When I came into school, I was proud to be Norman's sister."

Ed interjects, "A good example of that—"

"I'm *talking*, Edward," Joan reprimands, employing a glare that would stop many a younger brother dead in his tracks.

Ed is undeterred. "I have this wonderful example where I happen to come back to my grammar school years later and I'm trying to get some teacher I ran into in the hall to remember me. So she couldn't remember me and couldn't remember me, and then—'Oh, yes, now I remember. You're Norman's little brother.' "

They kick the topic of Norman around some more, and then Ed steps in with the definitive analysis. "Socially, he was advanced and I was backwards."

"No, I wouldn't say advanced and *back*wards," Joan says. "I never thought of you as backwards."

Ed won't take no for an answer. "What does it mean to be backwards?" he asks rhetorically. "I couldn't conduct a conversation with a girl or arrange for a date or get invited to a dance. I was not invited to a single party or dance throughout my whole high school time. Not once."

"Invited?" asks Macone, who, one gathers, doesn't remember seeing a lot of gold-embossed invitations during his adolescence.

Joan eagerly picks up this line of attack. "Edward," she explains, "those dances were held for just *people*."

"You were just supposed to show up," Joyce chimes in.

Ed replies: "What I mean is, they were—my *friends* had parties."

"Oh, okay," Macone says, now comprehending the depth of the tragedy: even the people Ed considered friends didn't want him at their parties. Not one to mince words, Macone observes, "Then they weren't your friends."

"Okay," Ed agrees, "I didn't have friends."

It seems strange for a man to win an argument with a line like this, but that appears to be what has happened. Now, secure in his victory, Ed can afford to admit that he had a few friends—"guys who were very much like me—"

Joan finishes the sentence—"science oriented." Having conceded now that Ed was not an avidly sought social commodity, Joan changes tack and argues that he didn't *want* to be one, anyway. "You may have had daydreams and desires and so on, to some extent, but your energy and your focus were elsewhere, Edward."

Ed disagrees. "I wanted to do the normal things, but I didn't know how. I was convinced there was something—"

Rather than complete the thought, presumably with the words *wrong with me*, Ed trots out the clinching anecdote. "When I was young, you know, sixth, seventh grade, two kids would be choosing sides for a game of something—it could be touch football. They'd choose everybody but me and then there'd be a fight as to whether one side would have to take me. One side would say, 'We have eight and you have seven,' and the other side would say, 'That's okay.' They'd be willing to play with seven."

Macone gets a big kick out of this.

Ed, of course, is not contending that he was the *only* social outcast in his school. "There was a socially active subgroup, probably not a majority; maybe forty percent were socially active. They went out on dates. They went to parties. They did this and they did that. The others were left out. And I was in this big, left-out group. But I was in the pole position. I was *really* left out."

Manuel Fredkin finally saved enough money to open a small radio parts store, thus expanding his son's already formidable potential for courting disaster in the name of science. One day, Ed took hundreds of army surplus batteries—the 45-volt type, bigger than a pack of cigarettes—and, after wiring them together, hooked one end of the series to a wooden stick and the other to a carbon rod. Slowly bringing the stick and rod near each other, he conjured up an electric arc, and, miraculously, did not kill himself.

By scraping off match heads and buying saltpeter, sulfur, and charcoal, Ed accumulated the ingredients for a good working knowledge of explosives. He built bombs not for their destructive power but for their aesthetic value; he liked the sight of a nice, healthy, four-foot mushroom cloud. Similarly, the rockets he fashioned out of cardboard tubing and aluminum foil were not instruments of aggression, though an observer could reasonably have mistaken them as such. One launching, he says, started a fire on his building's rooftop. Another vaporized the eyebrows and bangs of a neighborhood girl who had been given a seat too close to the launch pad. ("I had never seen her before, and I was very careful never to see her again.")

More than bombs, more than rockets, it was mechanisms that captured Ed's attention. From an early age he was viscerally attracted to Big Ben alarm clocks, which he methodically took apart and, conditions permitting, put back together. He also picked up his father's facility with radios and household appliances. But whereas Manuel seemed to fix things without deeply understanding them, Ed was curious about the underlying science. (He never joined that great high school institution of the 1940s, 50s, and 60s—radio club. "People in radio club wanted to talk to other people. That didn't interest me in the least. I was interested in what the electrons were doing.")

So Edward Fredkin—faced with a brother who was six years older, a bit distant, and nearly flawless; a sister who was, well, a *girl*; a father whose approval was always elusive; teachers who were not only boring but unjust; and classmates who, with a few exceptions, weren't on his wavelength and didn't care to be—spent much of his time manipulating mechanisms. While other kids were playing baseball, or chasing girls, or doing homework, Ed was taking things apart and putting them back together. Teachers were dull, toasters intriguing; children were aloof, even cruel, but a broken clock always responded gratefully to a healing hand. "I always got along well with machines," he remembers.

Lunch is over. The meal's final disagreement was about whether computers should ever serve as judges and juries (assuming they attain enough intelligence). Ed argued in the affirmative and Joan the negative. Now, an hour or so later, Joan is lying on Long Beach, along the island's south side, under a thatched sunshade, wearing a green two-piece bathing suit. She is fondly recalling the days when Ed, small for his age, could be subdued with her world-class scissors hold. (He deserved it; he would follow her around, softly echoing every word she uttered, with the aim, apparently, of driving her crazy.)

The two were not constantly at war. Indeed, for a while, at a very early age, they jointly explored the Big Questions. It was Joan who introduced Ed to the possibility that they, and everyone else on earth, were not "real," but part of a very long dream that God was having—an idea that, bizarre as it may sound, is not all that far removed from Fredkin's present thinking on the subject. And Joan vividly remembers jointly contemplating the paradox posed by two seemingly self-evident propositions: the universe must have an end, like everything else, but it would be impossible for *nothing* to exist. "We would be walking around," she says. "We'd go, it *has* to end . . . but it *can't*." One of her most cherished memories is of making mud pies with Edward, then in diapers, and of their mother approaching, tucking a mud-caked child under each arm, and heading back into the house.

Their mother died of cancer when Ed was ten. To this day, he is reluctant to talk about her. "I noticed that," Joan says when I bring it up. "It blows me away. I tell you, I think—I don't think deep inside he ever forgave her for dying. I think he's blocked an awful lot about Mother. I think she's very shadowy to him." Joan remembers her mother clearly. She was a warm, demonstrative woman, a reliable source of affection in a turbulent household. She had studied piano at a Russian conservatory, and in America her performances were sometimes broadcast on radio. When times were hard, she gave piano lessons.

Her death ramified for a long time. Within a year, Joan and Edward had to leave home, though neither is now clear on why. It apparently had to do with the difficulty of raising a family single-handedly while working, and Joan remembers something about an eviction notice. Whatever the rationale, Manuel told Joan to find someplace to live, and he arranged for Ed to stay with an aunt. "It was a dreadful situation I moved into—sort of like Oliver Twist," Joan says. "And Edward's was dreadful in other ways." He next was sent to live with another aunt, and then he spent some time with his married half sister, Hedda, his mother's daughter by a previous marriage. Only when his father remarried did Ed move back home.

Joan thinks that at some point during the interim Ed underwent a transformation; he resolved to secure a happy life, adversity notwithstanding. "I think he went through a very painful time and then he emerged with something intact and went on with things." He cultivated, she believes, an "impersonal intelligence," a detachment from day-to-day affairs that protected him from his own emotions. And, increasingly, he seemed to have confidence in his ideas, a confidence that could withstand even his father's doubts. By the time he graduated from high school, Joan says, "I don't think he needed validation."

THREE

A FINELY MOTTLED UNIVERSE

The prime mover of everything, the single principle that governs the universe, lies somewhere within a class of computer programs known as cellular automata, according to Ed Fredkin.

The cellular automaton was invented in the early 1950s by John von Neumann, one of the architects of computer science and a seminal thinker in several other fields. Von Neumann (who was stimulated in this and other inquiries by ideas of the mathematician Stanislaw Ulam) saw cellular automata as a way to study reproduction abstractly, but the word *cellular* is not meant biologically when used in this context. It refers, rather, to adjacent spaces—cells—that together form a pattern. These days, the cells typically appear on a computer screen, though Von Neumann, lacking this convenience, rendered them on paper.

In some respects, cellular automata resemble those splendid graphic displays produced by patriotic masses in authoritarian societies and by avid football fans on conservative American college campuses. Holding up large colored cards on cue, they can collectively generate portraits of Lenin, Mao Tse-tung, or a University of Southern California Trojan. More impressive still, one portrait can fade out, and another crystallize, in no time at all. Again and again, one frozen frame melts into another. It is a spectacular feat of precision and planning.

But suppose there were no planning. Suppose that instead of memorizing a long succession of cards to display, everyone learned a single rule for repeatedly determining which card was called for next. This rule might assume any of a number of forms. It could, for example, be designed to harness the collegiate preoccupation with peer-group behavior; in a crowd where all cards were either blue or white, each card holder could be instructed to look at his card and the cards of his four nearest neighbors—to his front, back, left, and right—and do what the majority did during the last frame. (This five-cell group is known as the von Neumann neighborhood.) We might call this the "1980s rule." The "1960s rule" might dictate that each card holder do the opposite of what the majority did. In either event, the result would be a series not of predetermined portraits but of more abstract, unpredicted patterns. If, by prior agreement, we began with a USC Trojan, its white face might dissolve into a sea of blue, as whitecaps drifted aimlessly across the stadium. Conversely, an ocean of randomness could yield islands of structure—not a Trojan, perhaps, but at least something that didn't look entirely accidental. It all depends on the original pattern of cells and the rule used to incrementally transform it.

This leaves room for abundant variety. There are many ways to define a neighborhood, and for any given neighborhood there are many possible rules, most a bit more complicated than blind conformity or unbending nonconformity. Each cell may, for instance, not merely count cells in the vicinity but pay attention to which particular cells are doing what. All told, the number of possible rules is an exponential function of the number of cells in a neighborhood; the von Neumann neighborhood alone has 2^{32}, or about four billion possible rules, and the nine-cell neighborhood that results from adding corner cells offers 2^{512} (about 1 with 154 zeros after it) possibilities. But whatever neighborhoods, and whatever rules, are programmed into the computer, two things are always true: all cells use the same rule to determine future behavior by reference to the past behaviors of neighbors, and all cells obey the rule simultaneously, time after time.

In the late 1950s, shortly after his acquaintance with cellular automata, Fredkin began playing around with rules, selecting the powerful and interesting and discarding the weak and bland. He found, for example, that any rule requiring all four of a cell's immediate neighbors to be lit up in order for it to be lit up at the next moment would not provide sustained entertainment; a single "off" cell would proliferate until darkness covered the screen. But equally simple rules could create great complexity. The first such rule discovered by Fredkin dictated that a cell be on if an odd number of cells in its von Neumann neighborhood had been on, and off otherwise. After "seeding" a good, powerful rule with an irregular landscape of off and on cells, Fredkin could watch rich patterns bloom, some freezing upon maturity, some eventually dissipating, some locking into a cycle of growth and decay. A colleague, after watching one of Fredkin's rules in action, suggested that he sell the program to a designer of Persian rugs.

Today new cellular automaton rules are formulated and explored by the "information mechanics group" founded by Fredkin at MIT's computer science laboratory. At the core of the group is an international trio—a physicist from France and two computer scientists, one from Italy and one from Canada. They differ in the degree to which they take Fredkin's theory of physics seriously, but all see some value in using cellular automata to simulate physical processes. In the basement of the computer science laboratory is the CAM—the cellular automata machine, designed by two of them (the Italian and the Canadian) partly for that purpose. Its screen has 65,536 cells, each of which can assume any of four colors and can change color sixty times per second. With this addition of two colors—an addition, incidentally, that makes the machine less reflective of Fredkin's theory—the number of rules for the von Neumann neighborhood grows from 2^{32} to 4^{1024}.

The CAM is an engrossing, potentially mesmerizing machine. Its four colors—the three primaries and black—intermix rapidly and intricately enough to form subtly shifting hues of almost any gradation; pretty waves of deep blue or magenta ebb and flow with fine fluidity and sometimes with rhythm, playing on the edge between chaos and order. One can imagine Timothy Leary spending an entire vacation within fifteen feet of the machine.

Guided by the right rule, the CAM can do a respectable imitation of pond water circularly rippling outward in deference to a descending pebble; or of bubbles forming at the bottom of a pot of boiling water; or of a snowflake blossoming from a seed of ice: step by step, a single ice crystal in the center of the screen unfolds into a full-fledged flake, a six-sided sheet of ice riddled symmetrically with dark pockets of mist. (It is easy to see how a cellular automaton can capture the principles thought to govern the growth of a snowflake: regions of vapor that find themselves in the vicinity of a budding snowflake freeze—*unless* so nearly enveloped by ice crystals that they cannot discharge enough heat to make room for new ice.)

These exercises are fun to watch, and they give you a sense of the cellular automaton's power, but Fredkin is not particularly interested in them. After all, a snowflake is not, at the visible level, *literally* a cellular automaton; an ice crystal is not a single, indivisible bit of information, like the cell that portrays it. But Fredkin believes that automata will more faithfully mirror reality as they are applied to its more fundamental levels and the rules needed to model the motion of molecules, atoms, electrons, and quarks are uncovered. And he believes that at the *most* fundamental level (whatever that turns out to be), the automaton will describe the physical world with perfect precision, because at that level the universe *is* a cellular automaton, in three dimensions—a crystalline lattice of interacting logic units, each one "deciding" zillions of times per second whether it will be off or on at the next point in time. The information thus produced, says Fredkin, is the fabric of reality, the stuff from which matter and energy are made. An electron, in Fredkin's universe, is nothing more than a pattern of information, and an orbiting electron is nothing more than that pattern moving. Indeed, even *this* motion is in some sense illusory: the bits of information that constitute the pattern never move, any more than football fans would change places to slide a USC Trojan four seats to the left. Each bit stays put and confines its activity to blinking on and off. "You see, I don't believe that there are objects like electrons and photons, and things which are themselves and nothing else," Fredkin says. "What I believe is that there's an information process, and the bits, when they're in certain configurations, behave like the thing we call the electron, or the hydrogen atom, or whatever."

The reader may now have a number of questions that will lead, unless satisfactorily answered, to major reservations about, if not outright contempt for, Ed Fredkin's theory of digital physics.

One such question concerns the way cellular automata chop space and time into little bits. Most conventional theories of physics reflect the intuition that reality is continuous—that one "point" in time is no such thing but, rather, flows seamlessly into the next; and that space, similarly, doesn't come in little chunks but, rather, is perfectly smooth. Fredkin's theory implies that both space and time have a graininess to them, and that the grains cannot be chopped up into smaller grains; it implies that people and dogs and trees and oceans, at rock bottom, are more like mosaics than like paintings, and that time's essence is better captured by a digital watch than by a grandfather clock.

The obvious question is: Why do space and time *seem* continuous if they are not? The obvious answer is: the cubes of space and points of time are very, very small; time seems continuous in just the way that movies seem to move when in fact they are frames; and the illusion of spatial continuity is akin to the emergence of smooth shades from the finely mottled surface of a newspaper photograph.

The obvious answer, it turns out, is not the whole answer. If Fredkin is right, the illusion of continuity is yet more deeply ingrained in our situation. Even if the ticks on the universal clock were, in some absolute sense, much slower than they are, time would still seem continuous to us, since our perception, itself proceeding in the same ticks, would be no more finely grained than the processes being perceived. So too with spatial perception: Can eyes composed of the smallest units in existence perceive those units? Could *any* informational process sense its ultimate constituents? The point is that the basic units of time and space in Fredkin's universe don't just *happen* to be imperceptibly small. So long as the creatures doing the perceiving are in that universe, the units *have* to be imperceptibly small.

Though some people may find this discreteness hard to grasp, Fredkin finds a grainy reality more sensible than a smooth one. If reality is truly continuous, as now envisioned by most physicists, then there must be quantities that cannot be expressed with a finite number of digits; the number representing the strength of an electromagnetic field, for example, could begin 5.23429847 and go on forever without ever falling into a pattern of repetition. That seems strange to Fredkin: Wouldn't you eventually get to a point, around the hundredth, or thousandth, or millionth decimal point, where you had hit the strength of the field right on the nose? Indeed, wouldn't you expect that *any* physical quantity has an *exactness* about it? Well, you may and may not. But Fredkin does expect exactness, and in his universe he gets it.

Fredkin has an interesting way of expressing his insistence that all physical quantities be rational. (A rational number is a number that can be expressed as a fraction—as a *ratio* of one integer to another. In decimal form, a rational number will either end, like 5/2 in the form of 2.5, or repeat itself endlessly, like 1/7 in the form of 0.142857142857142 . . .) He says he finds it hard to believe that a finite volume of space could *contain* an infinite amount of information. It is almost as if he views each parcel of space as having the digits describing it actually crammed into it. This seems an odd perspective, one that confuses the thing itself with the information representing it. But such an inversion between the realm of things and the realm of representation is common among those who work at the interface between physics and computer science. Contemplating the essence of information seems to affect the way you think.

The prospect of a discrete reality, however alien to the average person, is easier to fathom than the problem of the infinite regress, which is also raised by Fredkin's theory. The problem begins with the fact that information typically has a physical basis. Writing consists of ink; speech is composed of sound waves; even the computer's ephemeral bits and bytes are grounded in configurations of electrons. If the electrons are in turn made of information, then what is the information made of?

Asking questions such as these is not a good way to earn Fredkin's respect—especially when you're asking them for the fifth time in the course of a four-day stay on his island. A look of exasperation passes fleetingly over his face. "What I've tried to explain is that—and I hate to do this, because physicists are always doing this in an obnoxious way—is that the question implies you're missing a very important concept." He gives it one more try, two more tries, three, and eventually some of the fog between me and his view of the universe disappears. I begin to understand that this is a theory not just of physics but of metaphysics. When you disentangle the two—compare the physics to other theories of physics, and compare the metaphysics to other ideas about metaphysics—both sound less farfetched than when jumbled together as one. And, as a bonus, Fredkin's metaphysics leads to a kind of high-tech theology—to speculation about supreme beings and the purpose of life.

All this we will come to shortly. For now we can only ponder the short answer to the question of what Fredkin's universe is ultimately made of: "I've come to the conclusion," he says, "that the most concrete thing in the world is information."

CHAPTER

FOUR

THE MOMENT OF DISCOVERY

The entrepreneur in Ed Fredkin came out early. At age eleven he was knocking on doors and offering to fix toasters, clocks, radios, and lamps for a quarter. Next, enticed by comic-book promises of lavish prizes, he peddled magazine subscriptions door to door. Thus did he acquire a Sterno camping stove, which consisted basically of a place to put a can of Sterno. At age twelve, Ed was throwing copies of the *Los Angeles Daily News* from his bicycle into front yards. During high school he worked at the Hunley Theater on Hollywood Boulevard, taking tickets, cleaning toilets, climbing a ladder in the wee hours of the weekend to change the marquee's big plastic letters. He also worked as an actuarial clerk at the Occidental Life Insurance Company, where he encountered a relative of the computer—a key punch machine, used to enter data on IBM cards.

Ed's intellectual independence grew apace. Not content to gainsay only relatives and teachers, he set his sights on physicists. The logical place to begin was with Albert Einstein. Ed greatly admired Einstein for having "embarrassed" conventional physicists with the theory of relativity, but upon reading popular accounts of the theory, he concluded that Einstein suffered from confusion. He found a number of flaws in the "thought experiments" used to illustrate relativity. As it turned out, the flaws lay not in Einstein's thinking, but in the popularization of it. Nonetheless, they impressed Ed, because many of the popularizers were scientists themselves. "What I discovered," he says, "was that almost all physicists have misconceptions about these things." To this day, he derives manifest satisfaction from pointing out errors in the thinking of physicists. He likes, as he puts it, "to punch holes in their most sacred cows."

After graduation from high school in 1952, Ed headed for the California Institute of Technology with high hopes. Freed at last from uninspiring classes and inaccessible classmates, he could look forward to intellectual fulfillment and a more appreciative social environment. Or so he thought. The students at Caltech turned out to bear a disturbing resemblance to human beings he had observed elsewhere. "They were smart like me," he recalls, "but they had the full spectrum and distribution of social development." Ed remained on the fringes of society. He had a few buddies, who shared with him, in addition to alienation, a shortage of funds that further narrowed social options. They rode bicycles; they bodysurfed in the Pacific; they once hitchhiked up to San Francisco Bay. And every so often, having saved up twelve dollars, Ed would bicycle with a friend to the Glendale airport and purchase a flying lesson.

Fredkin's home in Brookline, Massachusetts, is bedecked with images of flight—a biplane suspended from the ceiling, a hand-carved wooden bird nearby, a rocket here, another plane there. Flying was Ed's chance to conspicuously establish his adulthood—indeed, his manhood; Norman had become an air force fighter pilot, but even Norman, Ed notes, hadn't flown a plane as early as age eighteen. Flying would eventually offer an escape from California and from the oppressive world of formal education. One of Fredkin's favorite cinematic moments is the point in *E.T.* when the children's bicycles are suddenly airborne.

In his case, the transition did not come so easily. Training flights were few and far between, and Caltech continued to be an unpleasant place. Ed had planned to major in physics, but his first lecturer was "a very old-fashioned guy" unduly concerned with the history of physics, and the small classes adjunct to lectures were taught by graduate students with no discernible enthusiasm for science. One of the few lessons Ed learned was that college is different from high school: if you don't study, you flunk out. This he did in 1954, a few months into his sophomore year.

The Korean truce was then barely a year old, and any young man not in college was draft material. Ed decided to exert some influence on his future and sign up before being signed up. Following in Norman's footsteps, he entered the U.S. Air Force Aviation Cadets training program, which began at Lackland Air Force Base in Texas and, over a year and a half of intensive training, continued in Mississippi and Arizona.

Ed thrived in cadet school. Sort of. "I didn't shine in the normal way. I was still a total social misfit as a cadet, exactly the same as in college. I still didn't really have a normal social life with girls, like other people did." Worse yet, he effortlessly acquired a reputation as something of a smartass; he had already learned, on his own time, a lot about subjects, such as aerodynamics and meteorology, that his classmates were confronting unprepared, and he didn't take pains to conceal his knowledge. Ed's saving grace in cadet school was that for the first time ever he played the game; he accepted the rules of an institution and wasted no time dwelling on their silliness. He absorbed the hazing—point-blank ridicule for, say, having substantial dust on his shoes—with composure. He avoided demerits. He even studied sometimes. And he easily adapted his flying skills to air force fighter planes. In 1955, he graduated.

Some of Ed's classmates didn't. Several, he recalls, died trying to master the nuances of flight—or, in some cases, trying to demonstrate mastery before acquiring it. One cadet talked so assuredly about his ability to fly at night through the gap between the Greenville Bridge and the Mississippi River that the only honorable thing to do was try. This and other deaths reinforced Fredkin's inordinate sense of caution and preparedness. He periodically revises his plans for surviving the nuclear war he believes to be imminent. At one point, Joyce says, he had concluded that by stationing two large guns atop Moskito Island, he could keep scavengers at bay in the wake of global holocaust. (Fredkin has been told that he was the model for Professor Steven Falken, a character in the 1983 movie *War Games*; Falken was a helicopter-flying computer scientist from MIT who lived on an island and despaired of averting nuclear war.)

At age twenty, Ed Fredkin was an officer in the air force and a thoroughly discontented young man. Not long after earning his cherished wings, he had been grounded because of recurring asthma. And his larger ambitions continued to elude him. He wanted badly to be at once normal and exceptional—to date and marry, like everyone else, and to perform intellectual feats like no one else. In fact, Fredkin says, he wanted these things too badly. An unnatural and repellent intensity came through when he spoke to women, and he worried so much about intellectual accomplishment that no time was left for it.

The thing to do, he finally decided, was reprogram himself. The ability to analyze and revise his behavior patterns had long been a source of pride. But in this case, the analysis of his predicament was the problem. How could it be the solution? "What I had to do was program myself to be unprogrammed, if that makes any sense—to be more relaxed." So he lowered his aims, tried to expect less of himself. And before long he was actually conducting sustained conversations with women. In 1956 he met one he especially liked, and he married her the next year. Meanwhile, some degree of physical coordination came to him. The first time he played golf, he recalls, he stepped up to a par-3 hole and, after hitting his first ball into the Great Beyond, teed up another one and hit a hole-in-one. "It was sort of like an ugly duckling growing up," he says of his final years in the air force.

A combination of rational thought, irrational thought, and randomness controls the fates of military men, and this is what finally brought Fredkin face to face with a computer. He was working in Florida at the Air Proving Ground Command, whose function was to ensure the quality of everything from combat boots to bombers. The command had recently been confronted with something that didn't fit naturally on that spectrum: SAGE. SAGE—Semi-Automatic Ground Environment—was a computerized air defense system that had been inspired by the Soviet Union's detonation of an atomic bomb in 1949. To test SAGE, the air force selected a group of men who knew little about computers (which, in fairness to the air force, is a description that applied to about everyone back then) and sent them to MIT's Lincoln Laboratory.

Some time after Fredkin and his colleagues arrived in Massachusetts, it became clear that SAGE, like virtually every computerized system before and since, was behind schedule. It could not be tested for another year. Meanwhile, about a dozen officers and enlisted men enrolled in computer science courses given by the contractors building SAGE. "Everything made instant sense to me," Fredkin remembers. "I just soaked it up like a sponge."

SAGE, when finally ready for testing, turned out to be even more complex than anticipated—too complex, indeed, to be tested by anyone but genuine experts; the job had to be contracted out. This development, combined with bureaucratic disorder, meant that Ed Fredkin was now a man without a function, a sort of visiting scholar at Lincoln Laboratory. "For a period of time that was probably over a year, no one ever came to me to tell me to do anything. Well, meanwhile, down the hall they installed the latest, most modern computer in the world—IBM's biggest, most powerful computer. So I just went down and started to program it." The computer was an XD-1. It had roughly the processing power of an Apple Macintosh and was roughly the size of a house.

When Fredkin talks about his year alone with this dinosaur, you half expect to hear violins start playing in the background. "My whole way of life was just waiting for the computer to come along," he says. "The computer was in essence just the perfect thing." He means this, apparently, in a nearly literal sense; the computer was preferable to every other conglomeration of matter he had encountered—more sophisticated, and more flexible, than other inorganic machines and more logical than organic machines. "See, when I write a program, if I write it correctly, it will work. If I'm dealing with a person, and I tell him something, and I tell him correctly, it may or may not work."

The speculation is too obvious to go unstated: Mightn't Fredkin's theory of digital physics be grounded in this odd affinity with digital machines? The computer, after all, is one of the first intelligent beings with which he was able to truly communicate. It is the ultimate embodiment of mechanical certainty, the refuge to which he retreated as a child from the incomprehensibly hostile world of humanity. Could the idea that the universe is a computer— and thus a friendly place, at least for Ed Fredkin—be wishful thinking?

This possibility has not escaped Fredkin's attention. Years ago, several of his students asked Philip Morrison, a physicist at MIT who has long been *Scientific American*'s book reviewer, what he thought of Fredkin's theory. Morrison's reply was reported to Fredkin as something like this: "Look, if Fredkin were a cheese merchant, he'd be telling you that everything in the universe is made out of cheese. But he happens to be a computer scientist, so he tells you that the universe is a computer."

Well, Fredkin admits, in a way Morrison is right; but in a way he's wrong. It has to do with the parallels between Edward Fredkin and Sigmund Freud.

Why, Fredkin asks, did the idea of the unconscious mind take so long to occur to someone? Because there are basically two kinds of people in the world: highly rational, scientific thinkers, who conceive of human behavior in mechanistic terms; and "flowery" thinkers— "humanists"—who hold the more romantic notion that behavior is guided by raw emotion and other irrational forces. Freud's theory fused these views; it entailed a schematic, mechanistic model of the mind, yet the model encompassed dark, stormy forces. So the theory's creation called for a rare thinker: a rational, scientific, flowery humanist. "The fact that someone finally did it is a near miracle," says Fredkin. "Why was it possible? Well, there's something in Freud's own psychological makeup that gave him some combination of being able to understand what was going on and to be a scientist." So too with Fredkin: by virtue of an extremely unusual intellectual history, he perceives a truth to which most people are constitutionally oblivious.

In a way, it makes sense. At a young age Fredkin acquired an interest in physics and forged a kinship with mechanism. Of course, many physicists develop such a kinship in youth, but in most cases it is diluted by formal education; quantum mechanics, the prevailing paradigm in contemporary physics, seems to imply that, at its core, reality has truly random elements, and thus lacks the mechanical predictability of a computer. But Fredkin, by dropping out of college, escaped this conclusion. (To this day he maintains, as did Albert Einstein, that it is based on a misinterpretation of the evidence. This is a critical belief, for if he is wrong, and the universe is *not* ultimately deterministic, then it cannot be governed by a process as exacting as computation.) After college, Fredkin joined the first generation of hackers and immersed himself in computer science, which led back to the study of physics, this time in the form of intensive self-instruction. For a time, he says, "there was no one in the world with the same interest in physics that had the intimate experience with computers that I did. I honestly think . . . that there was a period of many years when I was in a very unique position." Meanwhile, everyone else, handicapped by conventional education, was unable to see the truth. "I was going to say that it's fortuitous, but it's not. It's just the opposite. It's sort of too bad that circumstances have conspired to sort of keep things from them."

Upon leaving the air force, Fredkin stayed briefly at Lincoln Labs, programming the new IBM 709 and teaching others to program it. He had hoped to join there one of the world's first artificial intelligence research programs—the "pattern recognition" group, which was trying to teach machines to distinguish, say, an *A* from a *B*. To land this job, Fredkin had to impress Oliver Selfridge, a figure of some prominence in the early history of cognitive science. Fredkin decided to take a low-key approach to self-advertisement: he began the interview by admitting to Selfridge that, because he had never finished college, there were some gaps in his education. No, he remembers Selfridge replying, there weren't gaps in his education; his education was one long gap; a few isolated segments were filled in. "I was disappointed that he was not intelligent enough to see how smart I was," Fredkin says. He left Lincoln Labs a few months later.

The first person in a position of influence to recognize Fredkin's potential was Joseph Licklider, who then worked at Bolt, Beranek, and Newman, a consulting firm in the Boston area now known for its work in artificial intelligence. Licklider, an engineering psychologist, had been studying the perception of sound. Eager to learn about computers, he made a deal with BBN: he would continue to oversee psychoacoustics research while setting up a computer research department. This balancing act could be performed only with the help of someone who knew a lot about computers.

Ed Fredkin, meanwhile, was not looking for a place to work. He had decided to start his own business, and had already ordered a Librascope LGP-30, a computer about the size of an office desk. He was not sure exactly where the money for the Librascope was going to come from—presumably from one of the numerous companies for which he planned to do free-lance progamming. He listed prospective clients in alphabetical order, beginning with Bolt, Beranek, and Newman. Thus, he recalls, did he wind up talking to Joseph Licklider. Licklider remembers: "It was obvious to me he was very unusual and probably a genius, and the more I came to know him, the more I came to think that was not too elevated a description." He convinced Fredkin to let Bolt, Beranek, and Newman buy the LGP-30 and to come work for them.

At age twenty-four, Fredkin had at last found a mentor. Working with Licklider, he says, was "like going to college." Licklider says: "Ed worked almost continuously. It was hard to get him to go to sleep sometimes." A pattern emerged. Licklider would give Fredkin a problem to work on—say, figuring out how to get a computer to search a text in its memory for any given sequence of letters. Ed would retreat to his office and return twenty or thirty hours later with the solution—or, rather, with *a* solution; he often came back with the answer to a question that was of no interest to Licklider. Fredkin's focus was intense but undisciplined, and it tended to stray from a problem as soon as he was confident that, *in principle*, he understood the solution.

This intellectual wanderlust is one of Fredkin's most enduring and exasperating traits; just about everyone who knows him has a way of describing it. "He doesn't really work. He sort of fiddles." "Very often he has these great ideas and then does not have the discipline to cultivate the ideas." "There is a gap between the quality of the original ideas and what follows. There's an imbalance there." Fredkin is aware of his reputation. In self-parody, he once brought a cartoon to John Macone's attention. In it, a beaver and another forest animal are contemplating an immense man-made dam. The beaver is saying something like, "No, I didn't actually build it. But it's based on an idea of mine."

Licklider tried to sell Fredkin on the value of "packaging" his work: setting a realistic goal, pursuing it steadfastly, and then moving on to the next goal. Ed would have none of it. "He followed a kind of random course among the great ideas he had," says Licklider. For example: every car's license plate could be equipped with a transmitter that constantly signaled its location to receivers embedded in the streets. Then the police would always know where every car was. "The next day he came in and explained why socially that was a very bad idea. But I think he worked on it a whole day first."

Among the ideas that congealed in Fredkin's mind during his stay at BBN is the one that gave him his current reputation as—depending on whom you talk to—a thinker of great depth and rare insight, a source of interesting but reckless speculation, or a crackpot.

The idea that the universe is a computer was inspired partly by the idea of the universal computer. This is a cheap play on words that demands explanation. *Universal computer*, a term that can accurately be applied to everything from an IBM PC to a Cray supercomputer, has a technical, rigorous definition, but here a thumbnail definition will do: a universal computer can simulate any process that can be precisely described. Fredkin wasted no time in exploring this fact's implications. His DEC PDP-1 could easily simulate, say, two subatomic particles, one positively charged and one negatively charged, orbiting each other in accordance with the laws of electromagnetism. It was a pretty sight: two phosphor dots dancing, each etching a green trail that faded into yellow and then into darkness.

But the beauty that Fredkin perceived lay less in the pattern than in its underlying logic. A fairly simple programming rule—just a few lines of code—accounted for the fairly complex behavior of the dots. Fredkin had taken a little information and with it generated a lot of information. Indeed, in getting to know computers, Fredkin had discovered a language whose hallmark is just this sort of economy of information—the language of the algorithm.

An algorithm is a fixed procedure for converting input into output, for taking one body of information and turning it into another. For example, a program that takes any number it is given, squares it, and subtracts three is an algorithm. It may not sound like a very powerful algorithm, and, really, it isn't; by taking a 3 and turning it into a 6, it hasn't created much new information. But algorithms become more powerful with recursion. A *recursive* algorithm is an algorithm whose output is fed back into it as input. Thus, the algorithm that turned 3 into 6, if operating recursively, would continue, turning 6 into 33, then 33 into 1,086, then 1,086 into 1,179,393, and so on.

The power of recursive algorithms becomes especially vivid when they are used to simulate physical processes. To simulate the orbits of charged particles on the PDP-1, Fredkin wrote a program that took their velocities and positions at one point in time, computed those variables for the next point in time, then fed the new variables back into the algorithm to get newer variables—and so on, hundreds of times a second. The several steps in this algorithm, Fredkin recalls, were "very simple and very beautiful." It was in these orbiting phosphor dots that Fredkin first saw the appeal of his kind of universe—

a universe that proceeds tick by tick and dot by dot, a universe in which complexity boils down to rules of elementary simplicity.

His discovery of cellular automata a few years later permitted him to indulge more lavishly his taste for economy of information and strengthened his bond with the recursive algorithm. The patterns of automata are often all but impossible to describe with traditional mathematics, yet absurdly easy to express algorithmically. Nothing is so striking about a good cellular automaton as the contrast between the simplicity of the rule and the richness of its result.

We have all felt the attraction of this contrast. It accompanies the comprehension of any process, conceptual or physical, by which simplicity accommodates complexity. Simple solutions to complex problems, for example, make us feel good. The social engineer who designs uncomplicated legislation that will correct numerous social ills; the architect who eliminates several nagging design flaws by moving a single closet; the doctor who traces gastrointestinal, cardio-vascular, and respiratory ailments to a single, correctible cause—all feel the same kind of visceral, aesthetic satisfaction that filled the first caveman who *literally* killed two birds with one stone. So do mystery buffs. They bask in the satisfaction of discerning the single malign intent that lies behind diverse death and duplicity. (Imagine how empty you would feel upon reaching the end of a mystery novel if you found that each murder had its own peculiar logic.)

The climax of a search for scientific unity, like the climax of a search for narrative unity, is a moment of gratification, of joyous, even ecstatic, comprehension. It is a reward powerful enough to warrant the trouble of getting to; scientists, like mystery readers, may enjoy the landscape along the way, but few would begin the journey if it wasn't leading somewhere. As Alfred North Whitehead put it: scientists don't discover in order to know; they know in order to discover.

The moment of discovery not only reinforces the search for knowl-edge and inspires further research; it *directs* research. The unifying principle, upon its apprehension, can elicit a devotion that thereafter serves as a guiding light. The scientist, now enthralled by the prin-ciple's power, tries to expand that power. Somewhat like a Buddhist monk or a born-again Christian, he looks everywhere for manifesta-tions of, and affirmations of, his unity. It was the scientist in Ed

Fredkin who, upon seeing how a single programming rule could yield nearly infinite complexity, got excited about looking at physics in a new way and stayed excited. He spent much of the next few decades fleshing out his intuition.

FIVE

THE ROAD TO RICHES

After a few blissful years of exploring and enriching computer science at Bolt, Beranek, and Newman, Fredkin began to wonder whether he would ever get his due there. He, as much as anyone, had helped prepare the company for the computer revolution. He had recommended the purchase of the Digital Equipment Corporation's PDP-1, which turned out to be a seminal machine, and had gotten it up and running. It was at his suggestion that BBN had hired Marvin Minsky, John McCarthy, and other luminaries who gave the place a reputation for intellectual fertility. And what had he gotten for all this? "They started me out at fairly low pay. It went up fairly fast, but still, I got them into the computer business, and they never thought to give me any kind of equity, because I was like a junior person and it wasn't clear how important the computer business was." Fredkin made a proposal: BBN would create a new division, and he would run it. He wasn't asking for much—just an office and a budget and a computer, really. He didn't get it.

The ensuing resignation did not surprise Licklider. "I could tell that Ed was disappointed in the scope of projects undertaken at BBN. He would see them on a grander scale. I would try to argue, 'Hey, let's cut our teeth on this and *then* move on to bigger things.'" Ed wasn't biting. "He came in one day and said, 'Gosh, Lick, I really love working here, but I'm going to have to leave. I've been thinking about my plans for the future, and I want to make'—I don't remember how many millions of dollars, but it shook me—'and I want to do it in about four years.' And he did amass however many millions he said he would amass in the time he predicted, which impressed me considerably."

The road to riches began with Fredkin's fear of nuclear war. It was 1961, and Kennedy and Khrushchev were doing some saber rattling. Fredkin and a colleague at BBN, Roland Silver, were looking for an entrepreneurial adventure, and both felt that the place to spend World War III was the southern hemisphere. Brazil seemed like a nice spot. Its government was encouraging foreigners to set up shop, and its economy, on the brink of modernization, seemed ripe for computerization. Fredkin and Silver decided to learn Portuguese, head south, and start a consulting company called Information International. As they prepared for departure, though, President Jânio Quadros resigned unexpectedly, casting the country's political stability into doubt. The closer they got to leaving the States, the less romantic Brazil sounded.

Fredkin, now without income, moved his family into Silver's home, and the two men continued to look for a good joint venture. But it became clear that they weren't meant for each other. They mutually reinforced their procrastinatory tendencies, getting less done together than either did alone. Finally, Silver took a job with the Mitre Corporation, and Fredkin went his own way, taking the name Information International Incorporated with him. It was an impressive name, for a company with no assets and no clients, whose sole employee had never graduated from college.

Actually, Triple-I, as the company came to be called, wasn't entirely without assets. The incorporation papers had a space for capital value, so Fredkin gathered up his typewriter, any books with relevance to computers, and other miscellany, evaluated them, and declared a capitalization of $700. He then found a landlord trusting enough to lease office space without receiving rent in advance, and struck a consulting deal with the Digital Equipment Corporation that was tailored to his circumstances: He would bill them each Friday at five o'clock, and his check would be ready on Monday morning. DEC also sold him office furniture at bargain-basement rates. Perhaps his most important piece of furniture was a cot. He often worked through the night, and sometimes productivity dictated an hour-long nap.

After DEC had gotten Information International moving, the Woods Hole Oceanographic Institution directed it toward wealth. One of Woods Hole's experiments had run into a complication; underwater instruments had faithfully recorded the changing direction and strength of currents, but the information, encoded in tiny dots of light on 16-mm film, was inaccessible to the computers that were supposed to analyze it. Faced with the problem, Fredkin pondered it a while and then rented a 16-mm movie projector. He aimed the projector at a cathode ray tube. Normally, a beam would pass from the projector's bulb through one lens, through the film, through another lens, and out to the cathode ray tube. But light could travel equally well in the opposite direction; if he turned off the bulb, any blips of light on the CRT would send faint beams *into* the projector, through one lens, through the film, through the other lens, and toward the bulb. And if he replaced the bulb with a photomultiplier, which amplifies light and converts it into electrical impulses, a digital description of the beams' locations could, with appropriate circuitry, be channeled into a computer, which could, with appropriate programming, record any data represented by them on magnetic tape. That is what he did. The cathode ray tube shot a comprehensive series of beams into the projector. Those beams that succeeded in reaching the photomultiplier had located dots on the film. The exact locations of the dots were noted by the computer.

This contraption pleased the people at Woods Hole and led to a contract with Lincoln Labs. Lincoln was still doing work for the air force, and the air force wanted its computers to analyze radar information that, like the Woods Hole data, consisted of patterns of light on film. A makeshift information conversion machine earned Triple-I $10,000, and within a year the air force hired Fredkin to build equipment devoted to the task. The job paid $350,000—the equivalent today of around a million dollars and by a wide margin the biggest fee in the company's history. RCA, too, needed to turn visual patterns into digital data and paid Fredkin to automate the process. Other such jobs came along, and this sort of translation became Triple-I's staple. The company built "programmable film readers" that sold for half a million dollars each. By 1965, Triple-I's annual sales had reached a million dollars, and in 1967 the figure was $1.7 million. Early the next year Fredkin offered shares of the company to the public. "Basically I never paid myself very much and then it went public, and I was suddenly a paper millionaire."

He soon began cashing in his chips. First he bought a ranch in Colorado. Then one day he was thumbing through the want ads and saw that an island was for sale. Owning an island—that seemed like a neat idea. He paid roughly a million dollars in cash and stock in

1968, and Moskito Island is now worth close to five million, he figures. Today he owns only a few token shares in Information International, whose main technology is still related, if distantly, to the 16-mm projector he rented in 1963. The company makes machines that record on film the information from which printing plates are then made. *Time, Newsweek,* and *The Wall Street Journal* are among the periodicals produced with Triple-I technology.

In 1962, at the suggestion of the Defense Department's Advanced Research Projects Agency, MIT set up what would eventually become its Laboratory for Computer Science. It was then called Project MAC. (The acronym stood for both Machine-Aided Cognition and Multi-Access Computer.) Fredkin had connections to the project from the beginning. Licklider, who had left BBN for the Pentagon shortly after Fredkin's departure, was influential in earmarking federal money for MAC. Marvin Minsky—who would later serve on Information International's board of directors and who by the end of 1967 owned some of its stock—was centrally involved in MAC's inception. Fredkin had served on Project MAC's steering committee, and in 1966 he began discussing with Minsky the possibility of becoming a visiting professor at MIT. The idea of bringing a college dropout onto the faculty, Minsky recalls, was not as outlandish as it now sounds; computer science had become an academic discipline so suddenly that many of its leading lights possessed skimpy credentials. Nonetheless, Bob Fano, the director of Project MAC, opposed the appointment, insisting that Fredkin qualified as no more than a visiting lecturer. Fredkin wasn't interested in being a lecturer. He wanted to be a visiting *professor.* "It's a characteristic of Ed's," Fano observes, "that he never accepts anything but the top spot."

Fano prevailed, but late in 1968, after Licklider had come to MIT and replaced him as head of Project MAC, the idea surfaced again. Minsky and Licklider went to bat for Fredkin and convinced Louis Smullin, head of the electrical engineering department, that Fredkin was worth the gamble. "We were a growing department and we wanted exciting people," Smullin says. "And Ed was exciting." The question then arose: What course would he teach? Fredkin proposed a course on problem solving. Smullin asked for more detail: How, exactly, would the course tie in with electrical engineering? It wouldn't, Fredkin explained. It would just be about problem solving—all kinds of problems, all kinds of solutions.

Being a good problem solver is to Ed Fredkin what being the author of "The Christmas Song" is to Mel Torme—a primary source of self-esteem, something he does not hesitate to share with others. Broken computers, lame generators, and stalled automobiles almost seem to begin working upon his arrival at the scene, he says. But his skill extends beyond the mechanical realm, and the further removed it is, the prouder he seems to be of it. He tells of helping a student avoid the draft; of saving the life of a baby on an airplane; of solving the murder of a colleague and close friend, amassing so much evidence that the prosecutor was embarrassed into reopening the case. In a typical dinner conversation, Fredkin will flit from one solution to the next, often unearthing undetected problems expressly for the purpose: Los Angeles could save money by using its water resources more efficiently; balloon pilots could stay aloft longer if they had small nuclear reactors on board.

Of all the hypothetical problems faced by students in his problem-solving class, Fredkin's favorite was the Doctor's Dilemma. Suppose that a doctor has been visited by someone from outer space who, in a parting gesture of good will, granted him miraculous restorative powers. A single finger, brushed ever so lightly against any part of anyone's body, will cure any disease. Now, the question is this: What is to be done? By the end of the course, Fredkin's students knew how to handle a question so general: begin by defining the most desirable outcome; if it is attainable, pursue it. In this case, the most desirable outcome is obvious: cure all illness. Is it attainable? Natch.

First, in the world's major cities you build conveniently located railways, flanked on each side by six slender, round bars spaced a mere inch apart. The bars are positioned a few feet off the ground, so that a man sitting in a roller coaster car could let his ten fingers glide between them as he traveled down the track. The rest is easy. You take all the region's sick people and line them up, five to a spot, each with one finger between two bars. Well, says Fredkin, perfectly reasonable people-spacing parameters, along with a moderate rate of speed, would permit the healer to touch 360,000 patients an hour. He could be in and out of New Delhi in a day.

Fredkin devoted one lesson to appeals to authority: faced with a problem outside your expertise, find someone whose expertise it is within. For homework, students had to employ exactly that method. This may not have helped Fredkin's reputation among faculty members (at least one called him, puzzled, to ask whether it was all right to give out the answer), but his students, apparently, were impressed. A year later, when Fredkin was up for appointment to a full professorship, his department chairman called a group of them in to discuss his teaching. "I don't know what you've done to those students," Fredkin remembers Smullin telling him later. "Maybe you've hypnotized them."

Fredkin's faculty appointment also had support from on high; Licklider and Minsky made a pitch directly to Jerome Wiesner, then MIT's president. Licklider remembers: "We said, essentially, It's a very good bet—a better bet than you're going to get with academic credentials. Jerry got acquainted with Ed and could see that what we said about Ed's mind was true." At age thirty-four, Fredkin became one of the youngest full professors at MIT—and by far the youngest without a college degree.

The problem with Project MAC, Fredkin decided in 1971, upon succeeding Licklider as its director, was low morale—a lack of esprit de corps. The solution was simple: eliminate professors whose work was not a source of pride. He called a meeting at which faculty members reported on the status of their research and after which they passed judgment, by secret ballot, on the worth of each project. Of about twenty projects, he found, four were held in low esteem by almost everyone. Thus began the dicey task of easing people out of MAC. "There would be a junior faculty member, and I had to say to the department, 'You know, I have to talk him into leaving the lab.' And they would say, 'That probably means he shouldn't be at MIT.' And I'd say, 'You're probably right.'"

Periodic winnowing of this sort, Fredkin says, gave the laboratory a sense of purpose. It also satisfied Fredkin's pronounced dislike of small potatoes. "He was what I would call a 'big idea' administrator," says Licklider. "He wanted to make the place really great in three or four areas and didn't want to mess around with picayune projects."

Fredkin's own pet project, for example, was to found an international artificial intelligence laboratory that would include scientists from the Soviet Union. "My thesis was that such a laboratory was needed before the world would get to a state where certain countries would decide that AI was of strategic importance, at which point it would be too late to create such a laboratory. This has now happened."

This particular brainstorm was but one reflection of Fredkin's impulse to save the world—or, at the very least, large parts of it. He has toyed with the idea of removing ambiguity from diplomatic discourse by developing an international, rigorously defined language. During the Vietnam War, he founded the "Army to End the War," an organization of students who tried, in vain, to bring order and discipline to a movement he found disturbingly chaotic. And Fredkin's entrepreneurial energy, he likes to point out, has often been steered by concerns other than profit. Around 1981 he began helping a group of blacks acquire control of the CBS affiliate in Boston, Channel 7, and he served as its president for six transitional months. (True, he made around $10 million as the company's stock appreciated, but, he says, this was not his sole motivation.) More recently he has looked into the socially productive uses of personal computers in the Third World. And he continues (with the blessings of the American government, he stresses) to try to sell Soviet officials on the wisdom of buying lots of personal computers from the United States—a goal that, he argues, is in the interests of Americans and everyone else. "I personally don't want to be in the business of selling computers forever," he says of this venture. "What I'm trying to do is to get the ball rolling. And then it can be better handled by other companies." There could be some money in getting the ball rolling, he admits. But "so far the money's going the other way."

In the early 1980s, Fredkin—tired, presumably, of beating around the bush—taught a course at MIT on saving the world. The idea was to view the world as a giant computer and to write a program that, if methodically executed, would lead to peace and harmony—the "global algorithm," it was called. Along the way, an international police force would be formed and nations would surrender some of their autonomy to international tribunals. "It's a utopian idea," Fredkin concedes, but he adds with emphasis that it's not anything so simplistic as a formula for *instant* utopia. "This is a *series of steps* . . .

that gets you to utopia." If more people would take the plan seriously, he says, it could succeed. "I'll make this strange sort of arrogant statement that the reason people think my ideas aren't practical is that . . . they don't understand that if they would just sort of act like machines it would all work."

Wherever the blame lies, Fredkin has not yet saved the world. There is no global police force, diplomatic discourse remains fraught with ambiguity, and research in artificial intelligence is still mostly a national affair. Some people think the failure of that last project was the main reason Fredkin gave up his post as director of Project MAC only two years after assuming it. Fredkin himself just remembers getting bored with the job. Whatever the cause of his departure, he went from one grand design to another.

His next mission was to throw himself wholeheartedly into developing his theory of digital physics. "Ed would like to do, I think, one of the really great things," says Licklider. "I think he's always been looking for something like digital physics, which at least offers the possibility of a radical breakthrough in the sciences. He doesn't want to be one of the top thousand scientists. If he can't be one of the top ten or twenty, he'd just as soon not be one."

CHAPTER
SIX

THE BILLIARD BALL COMPUTER

Richard Feynman was regarded by some scientists as the smartest person of his generation. His lectures on physics were recorded in two massive volumes that have attained roughly the status reserved in other circles for the Bible. At least one colleague, quoting from them in an article, has seen fit to divide the sentences into verse. Among all of Feynman's fans, there is probably none more devoted than Ed Fredkin. Fredkin learned quantum mechanics at Feynman's knee, and Feynman's respect was one of the few seals of approval he earned from the physics establishment. Feynman was best man at Fredkin's second wedding, and Richard Fredkin is named after him.

The two men met in the early 1960s. Fredkin and Marvin Minsky, after traveling west on business, found themselves in Pasadena, home of the California Institute of Technology, with time to kill. "So we decided, hmmm, here we are in this place with all these great people and everything. Let's call a great person." Fredkin suggested Linus Pauling, but he wasn't home. Minsky suggested Feynman. "So we just called him out of the blue. He had never met either of us or heard of us, and he invited us over to his house and we had an amazing evening. I mean, we got there at like eight or something and stayed till three in the morning and discussed an amazing number of things." Fredkin kept in touch, and over the next few years he sometimes lodged at the Feynman house.

By the early 1970s, Fredkin had decided that he would have to learn more about quantum mechanics if his theory of digital physics was to progress. And who better to teach him than Richard Feynman? He asked Feynman if there was any way for an MIT professor to spend a semester or two at Caltech. Feynman exerted some influence (which he possessed in abundance, having won the Nobel Prize in 1965) and convinced the physics department to designate Fredkin a

Fairchild Distinguished Scholar. In the fall of 1974, Fredkin began a year-long sabbatical.

"The deal we had," Fredkin remembers, "was that I would teach him about computers and he would teach me about physics." Fredkin met sometimes with Feynman and sometimes with other members of the faculty, gradually acquiring, one of them recalls, a coterie of professors who were fascinated by his mind. What impressed them was the power of his naiveté. Lacking a formal education in math and physics, he confronted problems long ago dismissed as hopeless or pointless. And sometimes he prevailed—unconventionally—with his peculiar array of tools. Feynman once said of Fredkin, "He's extremely fertile with ideas, and many of them turn out to be good, although at first they don't appear that way, at least to me. I've had him start on some new project and could never understand why he would think that would be interesting, and at the end I find out that he had the right instincts."

Pressed for an example, Feynman cited the question of whether it is possible, in principle, to build a perfectly efficient computer—a computer that doesn't use up any energy and therefore doesn't give off any heat. This issue has no practical significance now, and it may never have any. Its context is an idealized universe, with no friction and none of the "noise" of randomly circulating molecules. But it is of immediate interest to Fredkin, because to prove that there is no minimum on the amount of energy a computer must dissipate, he had to design something called a "reversible" computer. And the theoretical possibility of such a computer is evidence—in Fredkin's mind, at least—that the universe *is* one.

Everything in the universe is reversible. This doesn't mean that you could take, say, a bunch of carbon dioxide molecules that are bouncing aimlessly off one another and induce them to turn around and retrace their steps. It means only that you could retrace their steps for them, if you took the trouble, because information about their past is implicit in their present.

Before you get confused, forget the molecules. Instead, imagine the textbook illustration of molecules in motion: billiard balls. Suppose that near the beginning of a game of eight ball—say, one twentieth of a second after the cue ball makes contact with the triangle of racked balls—you could somehow suspend time and measure the velocities and directions of all the balls. You could then, with pencil and paper, retrace their paths back to the rack. Indeed, even a second or so later, after some balls have caromed off several banks and one another, you could, with some difficulty, infer fairly precisely the history of all the balls. After all, if a ball comes off a bank at 37 degrees, it must have hit the bank at 37 degrees (assuming it has no sidespin, which is just the kind of thing you're allowed to assume in discussions like this).

So far as we know, this is the way reality is. If you could measure everything in the universe with absolute precision, and then do a series of ungodly calculations, you could reconstruct the past. It would be sort of like filming the final two seconds of a billiard shot, then running the film backward and, after reaching its beginning, extrapolating—producing an animated segment that follows the film's characters further back into time. Once you move from the billiard table to the universe at large, of course, such backward extrapolation is not practical, for a variety of reasons. But it is in principle doable. The information for reconstructing history is out there. The universe remembers.

Computers, generally speaking, don't. If asked what 2 + 2 is, a computer will tell you 4, but by the time it does, it may well have destroyed its record of the question; for all it knows, you asked it what 3 + 1 is. That is the way computers are built—to destroy information along the way, lest they get clogged up. So you can't infer the past state of a computer from its present state; computers are not reversible.

There is a contradiction here. On the one hand, we've said that the universe is reversible—that its past can, in principle, be inferred from its present. On the other hand, we've just found something in this universe—a computer—whose past cannot be inferred from its present even in principle. Something must be amiss; if the universe really remembers, then the information about the computer's history is lying around somewhere. But where?

In 1961, Rolf Landauer of IBM's Thomas J. Watson Laboratory addressed this question rigorously. To understand his answer—or, more realistically, to approach a vague understanding of it—think back to the billiard table. As you might imagine, it gets harder to piece together the history of the balls as time passes and they have bounced off many banks and many other balls and are beginning to roll to a halt. And once they are standing still, it is downright *impossible* to infer their past states from their present states, even approximately. Obviously, if you take a film of a ball that's standing still and run it backward, you'll have a film of a ball standing still; there will be nothing to extrapolate from. The billiard table thus finds itself in the position of the typical computer, lacking a record of the past. Again, the question arises: Where has the information gone?

Landauer concluded that the information has floated off into the billiard hall in the form of heat—heat generated by the friction and air resistance that dragged the balls to a halt. More precisely, the information is in the molecules—air molecules and green felt molecules, mainly—whose frenetic motion amounts to heat. They have absorbed the balls' velocity and, thus, information about the balls' past. If you could keep track of all these air and green felt molecules, measure their velocities and directions, and then visually trace their influence backward to the balls, you would see—in your animated extrapolation into the past—each motionless ball begin to move. And all along this reverse path, the ball would be accelerated by other agitated molecules returning to the source of their agitation and, with reverse agitation, imparting further force to the ball.

So really there *is* enough information to reconstruct the history of a motionless billiard ball, enough information to run its film backward. It's just that the information has been chopped up into little bits. The ball's environment—the green felt and the air—has taken the ball's history, which the ball was carrying around in fairly compact form, and spread it all over the place in the form of heat. This is why when physicists use a billiard table as a model of molecular interaction, they explicitly assume that neither the balls nor the table are susceptible to friction; that way the balls can move eternally and carry their histories with them, just like molecules.

With an argument too profound for me to comprehend, much less convey, Landauer established that what is true of billiard balls is true of computers: if information is lost, energy must be lost in the form of heat. Thus, the computer's 2 + 2 hasn't really been destroyed; it has just been discarded, and the card carrying it has been torn up into a million pieces and tossed off into the atmosphere, where it can be pieced together only with immense difficulty—and where, in the meanwhile, it will constitute heat. In the case of a conventional computer—an electronic computer, that is—the electrons representing 2 + 2 are zapped off into nearby space, where they stir things up, and thus heat things up, just a little.

In real life, of course, this discharge of information is not the *only* reason computers dissipate energy in the form of heat; the heat given off by a personal computer has mainly to do with other things. But those things are peculiar to its technology. They are a by-product of the fact that it happens to run on electricity and use resistors. Computers don't have to run on electricity. Granted, whatever else they might run on probably would, like electricity, create some heat that has nothing to do with the loss of information. But that's beside the point. The point is that as the technology is refined, the amount of heat created per computational step can fall without limit—*except* to the extent that the computers discard information; the heat resulting from information loss has an absolute minimum, independent of the technology employed. This minimum is set by basic physical laws impinging on the processing of every kind of information.

You may ask: What is that minimum? The answer is: kT per irreversible function, where k is Boltzman's constant and T is the temperature. There. Happy?

The issue of reversible computation well illustrates the weird world in which Fredkin lives, along with all the other people, such as Landauer, who work on the physics of computation. It is not just a world in which information, often thought of as an abstraction, is seen as physical, and therefore as subject to the laws of physics. It is also a world in which the reverse is true: *everything* physical—a bunch of molecules, a rack of billiard balls, a baseball, an outfielder, his glove—is an information-processing system, a running record of itself. Perhaps that is why Fredkin worries about that implication of modern physics that seems to bother almost no one else: the problem of "fitting" an infinite amount of information into a finite volume. And perhaps that is why it is so easy for him to think of matter as being made of information. After all, if everything is a record of itself, then what is the difference between the physical thing and a perfectly accurate simulation of it? This may at first seem like a dumb question, but the more you think about it, the less easy it is to dismiss. (Still, at some point you probably should dismiss it, if only to ensure that you don't wind up telling cab drivers, blind dates, and other casual acquaintances that the universe is a computer.)

Rolf Landauer probably deserves a spot in science's hall of fame for realizing that the question of whether the dissipation of energy is inherent in computation—of whether there is some floor on the amount of heat computers exude—comes down to the question of whether erasing information is itself inherent in computation. However, he apparently will not win the veneration of posterity for his initial answer to this question. He concluded that computers are indeed necessarily irreversible and thus necessarily dissipate energy, giving off heat. His logic—so seemingly solid that it was not challenged for a decade—went as follows. At the core of every computer on the market are lots of gates—notably "and" gates and "or" gates—that translate digital input into digital output. An "and" gate, for example, has two input lines and one output line, all of which can represent—through their level of voltage, typically—either 1 or 0. If the voltage in both input lines represents 1, the output line will then register a 1. But if a representation of 0 enters either input line, or both, the output line will register a 0. So, when the output line reads 0, there is no way of knowing exactly which representations the two input lines previously housed; information loss, and therefore energy dissipation, is inseparable from computation as we know it. The electrons representing information are routinely banished without a trace. And, while the heat they then constitute does "remember" the erased information, the computer itself has no recollection.

Strictly speaking, Landauer's contention—that, although the universe never forgets, computers always do—didn't contradict Fredkin's belief that the universe *is* a computer. It is conceivable that an irreversible process at the very core of reality could give rise to the reversible behavior of molecules, atoms, electrons, and the rest. After all, irreversible computers (that is, all computers on the market) can simulate reversible billiard balls. But they do so in a convoluted way, says Fredkin, and the connection between an irreversible substratum and a reversible stratum would, similarly, be tortuous—or, as he puts it, "aesthetically obnoxious." Fredkin prefers to think that the computer underlying reversible reality does its work gracefully. So at Caltech he set out to prove that computers don't *have* to destroy information—that a reversible computer is in principle possible.

He succeeded. He invented what has since become known as the "Fredkin gate." Instead of two input lines and one output line, it has three of each, and its input can always be inferred from its output. Fredkin showed that an entire computer could be built with such gates, and that, by using a special logic designed to conserve information, it could do anything any other computer can do. He had created—on paper, at least—a reversible universal computer.

Upon returning to MIT in 1975, though, Fredkin found that not everyone appreciated the importance of his gate, or of his work generally. "I was at that time a kind of whipping boy at MIT. I was spending my time doing this stuff, and they thought that it was all nonsense. . . . Now that it's gotten some publicity, and been written about in *Scientific American*, they would deny having said it was nonsense. But they made it very clear at the time."

In deeming the Fredkin gate nonsense, faculty members had their choice of rationales. The first was guilt by association: Ed Fredkin has this quasi-religious conviction that the universe is a computer, and the Fredkin gate is somehow tied in to this whole thing. The second was that Fredkin almost never publishes his ideas. Sometimes other people take the trouble to write them down and thus earn co-authorship. (In this case, a paper appeared in 1982, in the *International Journal of Theoretical Physics*: "Conservative Logic," by Fredkin and Tomasso Toffoli, the Italian in the information mechanics group.) But only once has Fredkin been the sole author of a published paper. So if his colleagues want to catch up on his recent intellectual feats, they have to rely on the grapevine, which few have the time or inclination to do. And if they want to dismiss his ideas out of hand, they can simply say that unless he wants to put them on the record, they're not worthy of consideration.

There was one other thing that made it easy to ignore the reversible computer: Fredkin still hadn't come up with a *clearly* reversible computer. It is one thing to describe a Fredkin gate and then argue abstractly that a suitable arrangement of such gates could do anything any other computer can do. It is another thing to actually design a nice, simple computer that clearly could work and clearly would not discard any information. This point was made with particular force by a doubter named Paul Penfield, a professor of electrical engineering at MIT. "This guy in some sense got on my case," Fredkin recalls. "So, I got mad and I decided I was going to find a simple physical realization."

Fredkin returned to the source, the very beginning of his need for a reversible computer: the fact that molecules, bouncing around like billiard balls, behave reversibly. What if, he asked, you designed a modified billiard table that could function as a computer? If you assumed that the balls would move eternally, and never slow to a halt under the drag of friction, how could anyone contend that the thing wasn't reversible? Thus was born the billiard ball computer. If it were ever actually built, it would consist of billiard balls caroming through a labyrinthine network of "mirrors," bouncing off the mirrors at 45 degrees, periodically banging into other balls at 90 degrees, and occasionally exiting through thresholds that occasionally would permit new balls to enter. To extract data from the machine, you would superimpose a grid over it; the presence or absence of a ball in a given square at a given point in time would, along with the direction of the ball, constitute information. Such a machine, Fredkin showed, qualified as a universal computer; it could do anything that more normal computers do. But unlike other computers, it would be perfectly reversible; to recover its history, all you would have to do is run it backward.

The billiard ball computer will never be built, because it is a Platonic device, existing only in a world of ideals. The balls are perfectly round and hard, and the table perfectly smooth and hard; there is no friction between the two, and no energy is lost when balls collide. Still, although these ideals are unreachable, through technological refinement they could be approached indefinitely, and the heat produced by friction and collision could thus be reduced without limit. Since no additional heat would be created by information loss, there would be no necessary minimum on the total heat coming from the computer. "The cleverer you are, the less heat it will generate," Fredkin says.

When the idea of the billiard ball computer first occurred to Fredkin, he did not conceal his glee. "I'm unlike other people. When they have a really good idea, they keep it a big, dark secret until they can explore all the easy consequences and get the credit for all of them. . . . I just told this idea to everyone I could find to try to interest them to work on it. I called Feynman up and told him the idea, and he just got it instantly on the telephone. This was the first sort of computer-like idea he really appreciated." A few days later Fredkin got a letter from Feynman with several ideas about how best to implement the billiard ball computer. To this day, one arrangement of "mirrors" in billiard-ball-computer architecture is known as the Feynman circuit. (It was independently invented by Andy Ressler, a student at MIT, but Fredkin decided to attach Feynman's name to it. "It's like a gift to Feynman," he says. "My idea was to keep him encouraged.")

If the billiard ball computer provided evidence that the universe is a computer—or, at least, that it *could be* a computer, or could *gracefully* be a computer—more such unconventional evidence was soon to follow. It came from Norman Margolus, the Canadian in the information mechanics group. Margolus showed how a two-state cellular automaton that was itself reversible could simulate the billiard ball computer using only a simple rule involving a small neighborhood. This cellular automaton in action looks like a jazzed-up version of the original video game: Pong. It is an overhead view of endlessly energetic balls caroming through clusters of mirrors and off each other. In a way, it is the best illustration yet of Fredkin's theory: it shows how a very simple, binary cellular automaton could give rise to the seemingly more complex behavior of microscopic particles bouncing off each other. And, as a kind of bonus, these particles themselves amount to a computer.

Though Margolus discovered this extremely powerful cellular automaton rule, it was Fredkin who had first concluded that it must exist and convinced Margolus to find it. "He has an intuitive idea of how things should be," says Margolus. "And often if he can't come up with a rational argument to convince you that it should be so, he'll sort of transfer his intuition to you."

That, really, is what Fredkin is trying to do when he argues that the universe is a computer. He cannot give you a single line of reasoning that leads inexorably, or even very plausibly, to this conclusion. He can tell you about the reversible computer; about Margolus's cellular automaton; about the many physical quantities, such as light, that were once thought to be continuous but are now considered discrete. And so on: the evidence consists of many little things—so many, and so little, that in the end he is forced, like the mystic, to convey his truth by simile. "I find the supporting evidence for my beliefs in ten thousand different places," he says. "And to me it's just totally overwhelming. It's like there's an animal I want to find. I've found his footprints. I've found his droppings. I've found the half-chewed food. I find pieces of his fur and so on. In every case, it fits one kind of animal, and it's not like any animal anyone's ever seen. People say, Where is this animal? I say, well, he was here; he's about this big, this that and the other, and I know a thousand things about him. I don't have him in hand, but I know he's there." The story changes upon retelling. One day it's Bigfoot that Fredkin's trailing. Another day it's a duck: feathers are everywhere, and the tracks are webbed. Whatever the animal, the moral of the story remains the same: "What I see is so compelling that it can't be a creature of my imagination."

The fervor of Fredkin's beliefs is hard to reconcile with the detachment that scientists are reputed to possess. But, as Margolus notes, science is not, in fact, a purely rational process. "People have these very strong prejudices about the way they think the world ought to be. And the ones who are right are the ones who make the great discoveries. You know, you can't just sit back and try all possible things. You have to have some sort of motivation for thinking this is the right way to go." Charles Bennett of IBM's Thomas J. Watson Laboratory, who independently arrived at a different proof that reversible computation is theoretically possible, makes a similar observation. The invention of the billiard ball computer (which Bennett considers more elegant in some ways than his own approach to the problem) is a textbook example of the power of conviction, he says. "Because he wanted so badly to find it, he found it."

But fervor has its drawbacks. By all accounts, it is very difficult to convince Fredkin that he's wrong, which, by all accounts, he is from time to time. And, while Fredkin sometimes inspires people, such as Margolus, to place his intuitions on a solid footing, sometimes he doesn't, in which case his intuitions remain private truths—removed from doubt in his mind, yet hidden from the scrutiny of science. He says, for example, that it would be easy to precisely simulate quantum mechanical phenomena, such as "the Stern-Gerlach apparatus," with a computer. Has he done it? No, but "there's no possibility of it being hard to do that," he asserts. Maybe not, but most scientists go ahead and conduct their experiments, just for good measure. And most scientists feel that their ideas can benefit from other people's ideas. Fredkin is a notoriously bad listener, and he seldom bothers to read the scientific literature.

In a way, his approach to physics is the scientific ideal turned inside out. He began by assuming that the universe is a computer, then figured out what smaller truths this large truth required; he actually compiled a list, around 1960, of things that would have to be true if indeed computation was the ultimate basis of physics. Then he began proving them. As he puts it, he "sort of embarked on a career of trying to make computers and physics like each other." This is not an illogical procedure, but it isn't science as popularly conceived, either.

According to popular conception, theories arise *in response* to evidence; data accumulate that no old theory can easily account for, and so a new one is born, of necessity. Granted, its creators may then, like Fredkin, seek data that consolidate its position. But ideally that evidence will have powers of discrimination; it will be not only consistent with the new theory but also inconsistent with competing theories. The evidence Fredkin has accumulated, while compatible with his theory, fits equally well with more conventional views of the world. There is nothing in the currently accepted laws of physics that forbids the construction of a reversible computer, or of a simple, reversible cellular automaton that can simulate such a computer.

As for the origin of Fredkin's theory: it arose not because it accounted for otherwise mysterious evidence, but because it seemed like a neat idea. This observation is not meant flippantly. Fredkin's theory seemed like a *literally* neat idea; it is intellectually clean and tidy—the kind of thing William of Ockham might have liked.

William of Ockham was an English philosopher who spent part of the early 1300s saying things like "Plurality is not to be assumed without necessity" and "What can be done with fewer is done in vain with more." His point, known variously as the principle of parsimony, the rule of economy of explanation, and Ockham's razor, was that if two competing theories can explain the same thing, the simpler theory should win. It is Ockham's razor, for example, that killed the pre-Copernican idea that the earth sits motionless at the center of the universe. The trouble with the geocentric theory wasn't that it couldn't explain the motion of the planets, but that it couldn't explain that motion as *simply* as the heliocentric theory; it offered less explanatory bang per buck.

Just about all scientists who have thought about the matter agree with William of Ockham that his razor is right. If you asked them why, most would refer to reality; nature has shown time and again that it really does operate simply. But is this attraction to simple and unified explanations really grounded in a detached appraisal of the history of science? Or are people just born with the principle of parsimony embedded in their brains? On first glance, at least, it seems that the latter is the case. Anyone who has felt what Ed Fredkin felt upon perceiving the unifying power of a cellular automaton (and all of us have—social engineers, architects, doctors, and mystery readers alike) has reason to suspect that he is experiencing something visceral; the Ockham epiphany has a fundamental feel to it.

Physics during this century has treated William of Ockham with both respect and contempt. The respect lies in the fact that various forces once considered "fundamental" have turned out to be different facets of *more* fundamental forces. For example, electricity and magnetism no longer require separate explanations; we now see that they are two sides of the same coin, a force called electromagnetism. Other sets of seemingly diverse forces have similarly been tied together in economical bundles. Indeed, physics appears to be heading toward a day when a single, ultimate force will account for all the forces once considered fundamental—not just electricity and magnetism, but also gravity and the forces responsible for the structure of molecules and atoms. The search for such a Grand Unified Theory occupied Einstein for the last part of his life and is now one of the hottest pursuits in science.

Fredkin is all for grand unified theories, and his theory of digital physics certainly qualifies as one. But he has a complaint about most current candidates for the job of unification: while deferent toward William of Ockham in their sheer power, and thus economy, they lack the simplicity of form that he probably would have liked. Differential equations, the language of much of modern physics, are alien to all who have not studied calculus—and to many who have. And, disconcertingly, the alienation grows as nature is more deeply penetrated; as we move from the laws governing the motion of planets to those governing atoms, electrons, and quarks, the mathematics becomes more arcane.

What Fredkin is saying is that William of Ockham should be commemorated not just in the number of things that each law accounts for, but in the form of the law. He finds it difficult to believe that nature, having generally possessed an artless elegance, is at heart a very baroque being. And if she *sounds* baroque when expressing herself in the language of physics, maybe the language of physics is the wrong language. Maybe nature's native tongue is the recursive algorithm, the language of the cellular automaton.

The rule underlying the typical cellular automaton could be fully comprehended by a bright third grader. With a pencil and a dozen reams of graph paper and enough time, he could trace the rule's influence through thousands of generations of cells, watching order emerge out of chaos, chaos engulf order, or both. If presented with differential equations describing these same processes, he would be nonplussed.

In many respects, the question of whether Fredkin is right about physics comes down to this difference between languages of description: Are algorithms more finely attuned to the texture of reality than differential equations? Fredkin thinks so. He shares William of Ockham's affinity for simplicity and believes that nature fundamentally does, too.

So one reasonable answer to the question sometimes asked by physicists—What can Ed Fredkin's theory of digital physics explain that older theories of physics can't?—is that it doesn't really *need* to explain anything new; it is by its very nature preferable to the old theories. If it can explain *just as much*, it should win on grounds of elegance. Well, then: Can Fredkin's theory explain *just as much* as the old theories?

The answer is no. Fredkin can show you how some differential equations can be translated into simple algorithms, and how a cellular automaton can create a pattern that looks something like microscopic particles bouncing around. But he hasn't come up with algorithms that account precisely for basic physical phenomena, much less the single rule that he believes governs the universe—"the cause and prime mover of everything."

Give him time, he says. Computers have only been around for a few decades, and the change of language he's talking about would be a scientific revolution on a par with Newton's classical mechanics, Einstein's theory of relativity, and quantum mechanics. Translating all the equations of physics into the language of computation represents a lot of man-hours, and there aren't many men on the job. "Look at quantum mechanics," he says. "Did one guy do it in his spare time in one year? No. It took an army of great men fifty years, and it's still going on." Fredkin counsels patience. "Someday we'll bridge that gap," he says. "There's no doubt eventually all that will be done."

CHAPTER
SEVEN
DEUS EX MACHINA

There was something bothersome about Isaac Newton's theory of gravitation. The idea that the sun exerts a pull on the earth, and vice versa, sounded vaguely supernatural and, in any event, was hard to explain. How, after all, could such "action at a distance" be realized? Did the earth look at the sun, estimate the distance, and consult the law of gravitation to determine where it should move and how fast? Newton sidestepped such questions; he fudged with the phrase *si esset*: two bodies, he wrote, behave *as if* impelled by a force inversely proportional to the square of their distance. Ever since Newton, physics has followed his example. Its "forces" and "fields" are, strictly speaking, metaphorical, and its laws purely descriptive. Physicists make no attempt to explain *why* things obey the laws of electromagnetism or of gravitation. The law is the law, and that's all there is to it.

Ed Fredkin refuses to accept authority so blindly. He posits not only laws but a law enforcement agency: a computer. Somewhere out there, he believes, is a machinelike thing that actually keeps our individual bits of space abiding by the rule of the universal cellular automaton.

With this belief Fredkin crosses the line between physics and metaphysics. The distinction between the two helps explain something that might otherwise seem puzzling: How can I say that the spirit of William of Ockham would smile on some bizarre theory about the universe being a computer? That doesn't sound like a very *simple* scenario, after all. True enough, but in comparing it with other theories of physics, we aren't concerned with the part about the computer. We are concerned only with the descriptive part—the idea that the dynamics of physical reality can be better captured by a single recursive algorithm than by differential equations, and that the continuity of time and space implicit in the traditional mathematics of physics is illusory.

As for the part about the computer, it should be compared only to metaphysical speculation about *why* traditional math seems to do a respectable job of describing reality. If some physicist were to claim that somewhere out there is a ministry of differential equations that sends subatomic police out to intimidate every particle into compliance with the law, that theory would be in competition with the theory that a computer animates the universe. The two ideas sound about equally ridiculous, so, metaphysically speaking, Fredkin's theory would be in a dead heat with its rival.

If Fredkin had Newton's knack for public relations—if he said only that the universe operates *as if* it were a computer—he could preserve the essence of his theory while improving his stature among physicists. In fact, some estimable physicists have lately been saying things not wholly unlike this stripped-down version of the theory. T. D. Lee, a Nobel laureate at Columbia University, has written at length about the possibility that time is discrete. And in 1984, *Scientific American*, not exactly a soapbox for cranks, published an article in which Stephen Wolfram, then at the Institute for Advanced Study, wrote, "Scientific laws are now being viewed as algorithms. . . . Physical systems are viewed as computational systems, processing information much the way computers do." In conclusion he declared: "A new paradigm has been born." (Wolfram, in his mid-twenties, is a MacArthur Foundation "genius" award winner and overall boy wonder who sometimes gets on Fredkin's nerves. He acquired recently the kind of interest in cellular automata that Fredkin has been pursuing for decades, and he is not, Fredkin says, always meticulous about crediting others for research on which his own ideas are based. As for Wolfram's opinion of Fredkin: "Ed has taken a bit more of an evangelical approach than is really optimal.")

The line between responsible scientific speculation and off-the-wall philosophical pronouncement was nicely illustrated by an article in which Tomasso Toffoli stayed just this side of it. Published in the journal *Physica D*, the article was called "Cellular Automata as an alternative to (rather than an approximation of) differential equations in modeling physics." Toffoli's thesis captured the core of Fredkin's theory yet had a perfectly reasonable ring to it. He simply suggested that the historical reliance of physicists on calculus may have been due not just to its merit, but to the fact that, before the computer, alternative languages of description were not practical.

Why does Fredkin refuse to do the expedient thing—leave out the part about the universe *being* a computer? One reason is that he considers reprehensible the failure of Newton, and of all physicists since, to back up their descriptions of nature with explanatory mechanisms. He is amazed to find "perfectly rational scientists" believing in "a form of mysticism: that things just happen because they happen." The best physics, Fredkin seems to believe, *is* metaphysics.

The trouble with metaphysics is its endless depth. For every question you answer, you raise at least one unanswered one, and it's not always clear that, on balance, you're making progress. For example, where is this computer that Fredkin keeps talking about? Is it in this universe, residing along some fifth or sixth dimension that renders it invisible, or in some metauniverse, or what? In some metauniverse, apparently. But that brings us to the question of the infinite regress. This question has been raised by Rolf Landauer, among others. Fredkin's theory reminds Landauer of the old turtle story. It is sometimes told about William James, but Landauer tells it about a fictitious venerable physicist. He has just finished lecturing at some august university about the origin and structure of the universe, and an old woman in tennis shoes walks up to the lectern. "Excuse me, sir," she says, "but you've got it all wrong. The truth is that the universe is sitting on the back of a huge turtle." The professor decides to humor her. "Oh, really?" he asks. "Well, tell me, what is the turtle standing on?" The lady has a ready reply: "Oh, it's standing on another turtle." The professor asks, "And what is *that* turtle standing on?" Without hesitation, she says: "Another turtle." The professor, still game, repeats his question. A look of impatience comes across the woman's face. She holds up her hand, stopping him in mid-sentence. "Save your breath, sonny," she says. "It's turtles all the way down."

The infinite regress afflicts Fredkin's theory in at least two ways. To begin with, if matter is made of information, what is the information made of? Ink? Teeny tiny radio waves? And even if we accept the contention that it is no less ludicrous for information to be the most fundamental stuff than for matter or energy to be the most fundamental stuff (and in a way, the three propositions *are* equally difficult to accept, when you really think about them), what about the computer itself? What is it made of? What energizes it? Who, or what, runs it, or set it in motion to begin with? These are questions for which Fredkin has an answer, but it is a subtle answer. Very subtle.

When Fredkin is discussing the problem of the infinite regress, his logic seems alternately cryptic, evasive, and vaguely appealing. At one point he says, "For everything in the world where you wonder 'What is it made out of?' the only thing I know of where the question doesn't have to be answered with anything else is for information." This puzzles me. Thousands of words later, I am still puzzled, and I press for clarification. He talks some more. And some more. And some more. What he means, as nearly as I can tell, is what follows.

First of all, it doesn't *matter* what the information is made of, or what kind of computer produces it. The computer could be a Play-doh Fun Factory, and some big-for-his-age six-year-old could be at the helm. Or, for that matter, we could all be inside the brain of a giant extraterrestrial (or, perhaps, supraterrestrial) octopus. What's the difference? Who cares what the information consists of? So long as the cellular automaton's rule is the same in all three cases, the patterns of information will be the same, and so will we, because the structure of our world depends on *pattern*, not on the pattern's substrate; a carbon atom, according to Fredkin, is a certain *configuration* of bits, not a certain *kind* of bits.

Besides, the question of what the information is made of has no practical importance, because we can never *know* what it is made of, or what kind of machine is processing it. This point is reminiscent of the childhood conversations between Ed and his sister, Joan, about the possibility that they were part of a dream God was having. "Say God is in a room and on his table he has some cookies and tea," Fredkin says. "And he's dreaming this whole universe up. Well, we can't reach out and get his cookies. They're not in our universe. See, our universe has bounds. There are some things in it and some things not."

The computer is not; hardware is beyond the grasp of its software. Imagine a vast computer program that contained bodies of information as complex as people, motivated by bodies of information as complex as ideas. These "people" would have no way of figuring out what kind of computer they owed their existence to, because everything they said, and everything they did—including formulate metaphysical hypotheses—would depend entirely on the programming rules and the original input. So long as these didn't change, the same metaphysical conclusions would be reached in a Kaypro 2 as in an Apple II, or, indeed, in a ten-billion-ton hydraulic computer made of sewage pipes and manhole covers.

This idea—that sentient beings could be numb to the texture of reality—has fascinated a number of people, including, lately, computer scientists. One source of the fascination is the fact that any universal computer can simulate another universal computer, and the simulated universal computer can, therefore, do the same thing. So it is possible to conceive of a nearly endless series of computers contained, like Russian dolls, in larger versions of themselves and yet oblivious to those containers. To anyone who has lived intimately with computers and thought deeply about their power, says Charles Bennett, this notion is very attractive. "And if you're too attracted to it, you're likely to part company with the physicists." Because physicists, Bennett notes, find heretical the notion that *anything* physical is impervious to experiment, removed from the reach of science.

Fredkin's belief in the limits of scientific knowledge may sound like evidence of humility, but in the end it affords great ambition; it helps him attack some of the grandest philosophical questions around. In fact, this broad power was one of the first things that attracted him to his theory. Long before he had found a reversible computer, even before he had encountered cellular automata, he realized that if our world was in some sense a simulation, several heretofore unanswerable questions could be answered.

For example, there is a paradox that crops up whenever people think about how the universe came to be. On the one hand, it must have had a beginning. After all, things usually do: people, movies, baseball games. Besides, the cosmological evidence suggests a beginning: the big bang. Yet science insists that it is impossible for something to come from nothing; the laws of thermodynamics forbid the amount of energy and mass in the universe from changing. So how could there have been a time when there was no universe, and thus no mass or energy?

Fredkin escapes from this paradox without breaking a sweat. Granted, he says, the laws of *our* universe don't permit something to come from nothing. But he can imagine laws that would permit such a thing. In fact, he can imagine *algorithmic* laws that would permit such a thing; the conservation of mass and energy is a consequence of *our* cellular automaton's rules, not a consequence of all such rules. Perhaps it is one of these more permissive cellular automata that governed the *creation* of our cellular automaton—just as the rules for *loading* software are different from the rules running the program once it has been loaded. Perhaps, in short, our universe was created by something whose creation is inherently not a mystery.

What's funny is how hard it is to doubt Fredkin when he so assuredly makes definitive statements about the creation of the universe—or when, for that matter, he looks you in the eye and tells you the universe is a computer. This is partly because, given the magnitude and intrinsic intractability of the questions he is addressing, his answers aren't all that bad. As ideas about the foundations of physics go, his are not completely out of the ballpark; as metaphysical and cosmogonical speculation goes, his isn't beyond the pale.

But there's more to it than that. Fredkin is, in his own odd way, a rhetorician of great skill. He talks softly, even coolly, but with a low-key power, a quiet and relentless confidence, a kind of high-tech fervor. And there is something disarming about his self-awareness. He's not one of these people who say crazy things without so much as a clue that you're sitting there thinking what crazy things they are. He is acutely conscious of his reputation; he knows that some scientists are reluctant to invite him to conferences for fear that he'll say embarrassing things. But he is not fazed by their doubts. "You know, I'm a reasonably smart person. I'm not the smartest person in the world, but I'm pretty smart—and I know that what I'm involved in makes perfect sense. A lot of people build up what might be called self-delusional systems, where they have this whole system that makes perfect sense to them, but no one else ever understands it or buys it. I don't think that's a major factor here, though others might disagree." It's hard to disagree when he so forthrightly offers you the chance.

Still, as he gets further from physics, and more deeply into philosophy, he begins to try your trust. For example, having tackled the question of what sort of process could give birth to a universe in which birth is impossible, he aims immediately for bigger game: *Why* was the universe created? Why is there something here instead of nothing?

When this subject comes up, we are sitting in the Fredkins' villa, a ten-minute walk, along a stone-and-cement path, from the hotel and restaurant. It is a nice spread, and probably justifies the $2,000 per week that people pay for it when the Fredkins aren't here. The living area has light rock walls, shiny-clean floors made of large white ceramic tiles, built-in blond wooden bookcases. There is lots of air—the ceiling slopes up in the middle to at least twenty feet—and the air keeps moving; some walls consist almost entirely of wooden shutters that, when open, let the sea breeze pass as fast as it will. I am glad of this. My skin, after three August days on the island, is charbroiled, and the air, though heavy, is cool; the sun is going down.

Fredkin, sitting on the white sofa, is talking about an interesting characteristic of some computer programs, including many cellular automata: there is no shortcut to finding out what they will lead to. This, indeed, is a basic difference between the "analytical" approach associated with traditional mathematics, including differential equations, and the "computational" approach associated with algorithms: analytically, you can predict a future state of a system without figuring out what states it will occupy between now and then; but in the case of a cellular automaton, you must go through all the intermediate states to get to the end—there is no way to predict the future except to watch it unfold.

This indeterminacy is very suggestive. It suggests, first of all, why so many "chaotic" phenomena, such as smoke rising from a cigarette, are so difficult to predict using conventional mathematics. (In fact, some scientists have taken to modeling chaotic systems with cellular automata.) To Fredkin, it also suggests that, even if human behavior is entirely determined, entirely inevitable, it may be unpredictable; there is room for "pseudo-free will" in a completely mechanistic universe. But on this particular evening Fredkin is interested mainly in cosmogony, in the implications of this indeterminacy for the big question: Why does this giant computer of a universe exist? It's simple, Fredkin explains: "The reason is there is no way to know the answer to some question any faster than what's going on."

I contemplate this statement for a few seconds and then ask if he wouldn't mind running it by me one more time. Perceiving that my confusion is fundamental, he takes another tack. Okay, he says, suppose, just for the sake of argument, that there is this all-powerful God. "And he's thinking of creating this universe. He's going to spend seven days on the job—this is totally allegorical—or six days on the job. Okay, now, if he's as all-powerful as you might imagine, he can say to himself, 'Wait a minute, why waste the time? I can create the whole thing, or I can just think about it for a minute and just realize what's going to happen so that I don't have to bother.' Now, ordinary physics says, well, yeah, you got an all-powerful God, he can probably do that. What I can say is—this is very interesting—I can say I don't care how powerful God is; he cannot know the answer to the question any faster than doing it. Now, he can have various ways of doing it, but he has to do every goddamn single step with every bit or he won't get the right answer. There's no shortcut."

Around sundown on Fredkin's island, all kinds of insects answer some sort of call to action and start chirping or buzzing or whirring. Meanwhile, the wind chimes hanging just outside the back door are tinkling with methodical randomness. All this music—eerie, vaguely mystical to begin with—is downright disorienting when combined with the extremely odd things Fredkin is starting to say. It is one of those moments, normally reserved for nightmares, when the context you've constructed falls apart and gives way to a new, considerably weirder, one. The old context, in this case, was that Ed Fredkin is an iconoclastic thinker who believes that space and time are discrete, that the laws of the universe are algorithms, and that the universe works *according to the same principles* as a computer. (In fact, he uses this very phrasing in his more circumspect moments.) The new context is that Ed Fredkin is this guy who sits around on an island in the Caribbean believing that the universe is very literally a computer—and that, moreover, it is being used by someone, or something, to solve a problem. It sounds like a good-news/bad-news joke:

the good news is that our lives have purpose; the bad news is that their purpose is to help some titanic hacker estimate pi to nine jillion decimal places.

Wondering if I have misunderstood, I press Fredkin for clarification. So, I ask, you're saying that the reason we're here is that there's some being who wanted to theorize about reality, and the only way he could test his theories was to create reality? "No, you see, my explanation is much more abstract. I don't imagine there is a being or anything. I'm just using that to talk to you about it. What I'm saying is that there is no way to know what the future is any faster than running this [the universe] to get to that [the future]. Therefore, what I'm assuming is that there is a question and there is an answer, okay? I don't make any assumptions about who has the question, who wants the answer, anything."

Okay, fine. But I still don't get it. If the universe is here because it's the most direct route to the solution of some computational problem, then there must be someone, or something, who, or that, set the thing in motion and is waiting to see what will happen, or died while waiting, or, after watching us on TV for a few billion years, got bored and went next door to visit the Coneheads, or something. Right? The more we talk, the closer Fredkin comes to the religious undercurrents he's trying to avoid. "Every astrophysical phenomenon that's going on is always assumed to be just accident," he says. "To me this is a fairly arrogant position, in that intelligence, and computation, which includes intelligence in my view, is a much more universal thing than people think. It's hard for me to believe that everything out there is just an accident." This sounds awfully like the position of Pope John Paul II and Billy Graham, and I convey this to Fredkin. He responds, "I guess what I'm saying is: I don't have any religious belief. I don't believe that there is a God. I don't believe in Christianity or Judaism or anything like that, okay? I'm not an atheist. . . . I'm not an agnostic. . . . I'm just in a simple state. I don't know what there is or might be. . . . But on the other hand, what I can say is that it seems likely to me that this particular universe we have is a consequence of something which I would call intelligent." You mean that there's something out there that wanted to get the answer to a question? "Yeah." Something that set up the universe to see what would happen? "In some way, yes."

My conspicuous skepticism still bothers him. Look, he says, suppose you were walking along in a desert and came across a machine with four wheels, an engine, a transmission, a dashboard, and all the rest—something that bore a remarkable resemblance to a car. Wouldn't you be safe in concluding that it was built by people for the purpose of getting from one place to another? Well, suppose you came across a machine that had a sign that said THIS IS A PROBLEM-SOLVING MACHINE and was whirring away. Wouldn't you be safe in assuming that it had been set up to solve some problem and was now doing that?

The universe doesn't come with a sign attached, I reply, stalling for time. And then I come up with a better counterargument. When I was about four years old, I saved every sharp-edged, triangular rock I came across, confident that they were arrowheads, left over from cowboy-and-Indian days. In one sense, I was right to call them arrowheads; they could have performed that function adequately. But I was ultimately wrong; piercing flesh was never their purpose. They were just rocks, no more the product of design than the dirt beneath them. Similarly, it is true that we have a universe well suited to the function of finding out what the future will be, but that doesn't mean it was created for that purpose.

It is obvious from his expression that Fredkin is not bowled over by this line of reasoning. But I am very proud of it. I am just beginning to imagine future philosophers winning tenure on the basis of their astute analyses of "Wright's famous arrowhead argument" when Fredkin breaks in with his comeback. It's not just that the universe is finding out the future, he says; it's finding out the future by the only method through which the future can be found out. I don't see what difference that makes, and Fredkin, rather than press the point, resorts to the Socratic method. He asks, "What are computers used for in this world?"

"To compute," I say.

"What do you mean, 'To compute'? To compute what?"

This would be a good time to break his momentum with defensive tactics—answering "Asparagus," for example. But I have long had an inexplicable desire to please professors, so instead I say, "Answers to questions."

He is pleased. "Right, every computer we have is to compute answers to questions. And I'm saying here's the biggest computer that anybody ever saw. I'm saying its purpose is to compute answers to questions."

"But that's a fundamentally teleological view of the universe," I say. This accusation will stop most scientists cold, and often induce retreat, but Fredkin puts the ball right back in my court: "What do you mean by that exactly?" I'm not sure whether he disagrees with me or simply doesn't know what the word *teleological* means. It wouldn't surprise me if the latter was the case: he is generally insensitive to the unwritten rules of scientific conduct, one of which is to scrupulously avoid even the faintest teleological overtones. In either event, his question forces me to realize that I'm not sure what the word means myself. And it is important to be clear on that, because many, if not most, attempts to attribute any sort of meaning or purpose to life involve teleology of one sort or another.

Thinkers of a teleological bent have a peculiar habit. When they hear the word *why*, they think of the future. For example: Why does dropped toast seem always to land jelly-side-down? Most of us, and certainly the scientists among us, would seek the answer in the stretch of time *preceding* the toast's landing: perhaps the laws of aerodynamics dictate that the toast stabilize in mid-descent with the heavier side down. But someone teleologically inclined would answer the question by reference to the stretch of time *after* the landing: the toast ends up jelly-side-down *so that* your day will get off to a terrible start.

Dropped toast is a bad example, really. Teleological thinkers are a fairly roseate group, and they prefer questions such as "Why does the rain fall?"—to which they can reply, "So that the beauteous flowers will grow." Or, "Why did the brains of humans evolve to such a high level of complexity?" "So that there would be literature and art and large museums." Whatever the question, and whatever the answer, the principle remains the same: according to the scientific outlook, life is the necessary result of antecedent causes, and history is thus a pushing process; according to teleology, life is the necessary antecedent of some ultimate goal, and history is a pulling process.

Actually, this is a fairly strict definition of teleology. In common use, the word is just loosely associated with purpose. That, in fact, is why it popped into my mind when Fredkin started talking about the purpose of the universe. But when he pressed me for a definition, I realized that he had come up with a good example of how the universe could have a purpose without history being a pulling process. The purpose of the universe, he claims, is to reach some final state, the solution to some problem, but the final state is not—indeed, cannot be—specified in advance. Of course, even granting life this modest degree of significance is enough to incur the suspicion of many scientists, but it's probably not enough to sustain charges of heresy; the idea of God as some prime mover, who wound up the universe and then let it run, and has kept his, her, or its hands off ever since, is not irreconcilably at odds with a scientific world view. At worst, Fredkin could be convicted of one of the lesser degrees of teleology.

CHAPTER

EIGHT

THE MEANING OF LIFE

On January 7, 1977, Ed Fredkin was sitting in an airliner on the San Juan runway, hoping that the Eastern Airlines computer had not seated anyone next to him. It had—Joyce. What struck him first was how irrationally she was dressed. It was 85 degrees outside, and she was wearing a full-length winter coat. She had a perfectly good explanation. After buying it at a thrift store in Massachusetts, she had brought it home to show to her mother, who subsequently filled the suitcase space reserved for it with baked goods.

Joyce, then twenty, was crossing a threshold. She had just settled some long-festering issues with her father, finally extracting some acknowledgment of her autonomy, and was on the verge of graduation from Bentley College. This left her with lots of decisions to make, and Fredkin, being a successful businessman, was well positioned to give advice, which he did. A few days later Joyce told her girl friend about the nice man who had talked with her all the way from San Juan to Boston and given her a ride home in his station wagon. *Station wagon?* "My girl friend said, Aha, a wife, three kids, and a dog. And I said, Gee, I never thought of that—plus, I'm not interested in him that way." Nonetheless, "next time I met him I said, I have questions to ask you, three questions. I said, One, are you married? He said, Yes. Two, are there kids? Yes. Three, what are their ages? Eighteen, sixteen, and fourteen, or something like that. And then I said, Then what are you doing here with me? And he said, I don't know." This answer made a favorable impression on Joyce. "He was honest. He didn't say he wasn't married or anything like that."

The ensuing divorce took half of Fredkin's assets. This left him with millions of dollars, but marriage to Joyce wasn't an exercise in frugality. In addition to her appetite for travel and clothes, there was the island. It had never shown a profit and had been closed for years, but Fredkin reopened it and handed its management over to his new wife, who was fascinated by the idea of turning it into a money-maker. The last time I talked to her, she still was, because doing it was still a mystery. "If it weren't for this property," she said, "we would be ex*treme*ly wealthy."

A few years into the marriage, Fredkin invested about three quarters of a million dollars in the Three Rivers Computer Corporation, a financially needy start-up firm that was building a new computer for engineers, and became chairman of its board. With that much at stake, he began working very long weeks and virtually living in Pittsburgh, the company's home. (While there he served as visiting professor at Carnegie-Mellon University, delivering six lectures on such subjects as robotics, digital physics, and the algorithm to save the world; it was the only way Joyce could qualify for admission to two graduate business courses she wanted to take.) Fredkin plucked Three Rivers from the brink of ruin, but this proved only a reprieve; the company died an agonizing death.

Lest his assets fall into the low millions, Fredkin continues to work on a variety of commercial enterprises. He has invested time and money in Encore Computers, which makes supercomputers. He consults for the Carnegie Group, an artificial-intelligence company. And he occasionally goes to Russia to talk about selling personal computers. His theory sits in the closet, collecting dust.

This is not the way Fredkin planned things. In 1974, upon returning to MIT from Caltech, he was primed to revolutionize science. Having done the broad, conceptual work (concluding that the universe is a computer), he would leave it for others to take care of the details—to translate the differential equations of physics into algorithms; to experiment with cellular automaton rules and glean the most elegant; and, eventually, to discover *the rule*, the single law that governs every bit of space and accounts for everything. "He figured that all he needed was some people who knew physics, and that it would all be easy," Margolus recalls.

One early obstacle was Fredkin's reputation. "I would find a brilliant student, he'd get turned on to this stuff and start to work on it. And then he would come to me and say, I'm going to work on something else. And I would say, Why? And I had a few very honest ones, and they would say, Well, I've been talking to my friends about this and they say I'm totally crazy to work on it. It'll ruin my career. . . . I'll be tainted forever." Such fears were not entirely unfounded. Fredkin is one of those people who arouse either affection, admiration, and respect, or dislike and suspicion. The latter reaction has come from a number of professors at MIT, particularly those who put a premium on formal credentials, proper academic conduct, and not sounding like a crackpot. Fredkin was never oblivious to the complaints that his work wasn't "worthy of MIT," nor to the movements, periodically afoot, to sever, or at least weaken, his ties to the university. Neither were his graduate students.

Fredkin's critics finally got their way. In the early 1980s, while he was president of Boston's CBS TV affiliate, someone noticed that he wasn't spending much time around MIT and dredged up a university rule limiting outside professional activities. Fredkin was finding MIT "less and less interesting" anyway, so he agreed to be designated an adjunct professor. As he recalls the deal, he was going to do a moderate amount of teaching and be paid an "appropriate" salary. But he found the size of the checks insulting, declined payment, and never got around to teaching. Not surprisingly, he was not reappointed adjunct professor when his term expired in 1986. His duties as head of the information mechanics group—which he had for years discharged only sporadically—were formally given to Toffoli, who had been serving as the group's de facto director.

Fredkin despairs, these days, of vindication. He believes that most physicists are so immersed in their kind of mathematics, and so uncomprehending of computation, as to be incapable of grasping the truth. Imagine, he says, that some twentieth-century time traveler visited Italy in the early seventeenth century and tried to reformulate Galileo's ideas in terms of calculus. Although it would be a vastly more powerful language of description than the old one, conveying its importance to the average scientist would be nearly impossible.

There are times when Fredkin breaks through the language barrier, but they are few and far between. He can sell *one* person on *one* idea, another on another, but nobody seems to get the big picture. It's like a painting of a horse in a meadow. "Everyone else only looks at it with a microscope, and they say, 'Aha, over here I see a little brown pigment. And over here I see a little green pigment.' Okay. Well, I see a horse."

There is a glimmer of hope. Fredkin's attempt "to make computers and physics like each other" has succeeded, if not exactly in the way he originally intended. Comparing a computer's workings and the laws of physics turned out to be a good way to figure out how to build a very efficient computer—one that harnesses the laws of physics with great economy. Thus have Toffoli and Margolus designed an inexpensive but powerful cellular automata machine, the CAM 6. The "machine" is actually a circuit board that, when inserted into a desktop computer, permits it to orchestrate visual complexity at a speed that can be matched only by general-purpose computers costing hundreds of thousands of dollars. Since the circuit board costs only around fifteen hundred dollars, this engrossing machine may well entice young scientific revolutionaries into joining the quest for The Rule. Fredkin speaks of this possibility in almost biblical terms. "The big hope is that there will arise somewhere someone who will have some new, brilliant ideas," he says. "And I think this machine will have a dramatic effect on the probability of that happening."

But even if it does happen, it will not ensure Fredkin a place in scientific history. After all, he is not really *on record* as believing that the universe is a computer. Although some of his tamer insights have been adopted, fleshed out, and published by Toffoli or Margolus, sometimes in collaboration with him, the closest thing to a published version of the theory of digital physics is the book *Calculating Space*, by Konrad Zuse, a German computer scientist whose parallel thinking on the subject did not come to Fredkin's attention until the late sixties (at which time he had the book translated into English).

Fredkin's rationale for not publishing has to do with, of all things, lack of ambition. He's just "not terribly interested. A lot of people are fantastically motivated by publishing. It's part of a whole thing of getting ahead in the world." Margolus has another explanation: "Writing something down in good form takes a lot of time. And usually by the time he's done with the first or second draft, he has another wonderful idea that he's off on."

These two theories have merit, but so does a third: Fredkin doesn't write for academic journals because he doesn't know how. His erratic, hybrid education has left him with a mixture of terminology that physicists don't recognize as their native tongue. Further, he is not schooled in the rules of scientific discourse. He seems unaware of the teleology taboo, and just barely aware of the line between scientific hypothesis and philosophical speculation. He is not politic enough to confine his argument to its essence: that time and space are discrete, and the state of every point in space at any point in time is determined by a single algorithm. In short, the same odd background that allowed Fredkin to see the universe as a computer prevents him from sharing his vision. If he could talk like other physicists, he might see only the things they see.

It's my last morning on the island. Ed and I are sitting in the restaurant, on a thin stretch of floor between the end of the bar and the ocean, engaging in a psychological struggle: we have conflicting ideas about how to spend our final conversation.

For his part, Ed is trying to make sure I don't leave the island with the wrong impression. He's been thinking about that lunchtime conversation, and he wants to stress that his childhood was more than domestic combat and social isolation. Arguing with his father was actually sort of fun—"like a sport." Granted, "my father would sometimes get very competitive in this thing—he had to win no matter what. But there was a very positive aspect to it, too. It was a great intellectual stimulus." And his father often challenged him with mathematical puzzles: "The teakettle was twice as old as the pot was when the pot was seven years older than the teakettle is now"—that kind of thing.

For my part, I'm trying to extract from Ed Fredkin the Great Quote—the kind of pithy self-summary or unwittingly symbolic observation that you could use to end a story about a man who thinks the universe is a computer.

There is a subplot, too, a more subtle tension between Ed and me. It has to do with the dark clouds approaching from the eastern horizon. My flight from Virgin Gorda to San Juan leaves in a few hours, and if things get stormy, Ed won't be able to fly me to Virgin Gorda. I could stay another night, but at $130 a pop, I'd just as soon not, even though meals are included. So I'm hoping the storm will blow over. Fredkin is hoping it won't—not so he can spend another day with me (he was delighted to hear last night that I'm almost out of cassettes) but because the island is parched; the water tank is running low.

My questions, meanwhile, are failing. The Great Quote is nowhere to be found.

I ask him if there's any connection between thinking big thoughts (such as, the universe is a computer) and doing big things (such as starting a company). Well, yeah, sort of, in a way, he guesses.

Who are Ed's heroes? At various times: Einstein, Bertrand Russell, Leibniz, FDR.

Would he trade all his money to see his ideas vindicated? "What you're asking me is would I trade my material wealth, which doesn't mean a lot to me, for the rest of the world knowing about my ideas what I already know about them?" He's right: it was a dumb question.

I've decided to go after the teleology angle. This is my natural inclination, really; I find teleology tempting. I have a basically scientific world view, and it seems to basically work, but it isn't, by itself, very reassuring. Personally, I don't like the idea that we're mere specks in a universe indifferent to our fates. Any hint that life has some meaning, evolution some purpose, would be refreshing.

Correct me if I'm wrong, I say, but "in your view, the reason all of this has happened, the reason you own this island, everything, is because something set out to solve a problem by simulation—right?" Well, he replies, it's misleading to focus on him, or on any one person, or on people generally. "I think that we're probably incidental to the problem and the solution, because you have to look at where did it put its resources? It put its resources in the galaxies and stars."

In one sense, he's right. There is more sheer energy tied up in stars than in humans. But aren't humans a little more complex than stars? Wouldn't it be easier for the great hacker in the sky to find an algorithm that simulated stars burning than to find one that simulated, say, the greater Los Angeles area at rush hour? In other words, aren't there more *conceptual* resources tied up in humans than in stars? With this idea vaguely in mind, I say, "But in a sense one of the most interesting things to happen is us."

Ed doesn't find my observation thought provoking. "So we say," he says. "There may be more interesting systems than you know. Who knows what's going on inside a star? . . . The idea that everything we see there is a total, natural, random piece of junk, and we're the only ordered stuff around, is very farfetched. So I believe that there's something much more complex about stars and galaxies than is recognized."

How uplifting: there's little, if any, real purpose in our lives, but there may be a lot inside the sun.

Here's my last attempt to find philosophical contentment through Ed Fredkin: Doesn't he find it interesting that now, billions of years after the birth of the universe, we have come along and created these machines that are little replicas of it? Isn't there something a little *weird* about that? Doesn't it suggest that someone or something *planned* this whole thing, rather than just setting it in motion with no clear idea of what would happen?

No dice. Fredkin doesn't believe that this is the first time computers have turned up on earth; life is itself a kind of computer. DNA works much like a cellular automaton, and pretty much everything between DNA and the computer also runs on information. "Now, the nervous system of an animal is another information-processing thing, just the nervous system that sends messages back and forth for simple kinds of sensations and so on. That's the second level. And then the intellectual level is yet another level. And then with our intellect, when we make books and paper and do algebra and so on, that's yet another level of information processing. The computer is yet another level. And then within the computer we get different levels." Moreover, he adds, society as a whole is an information-processing system. A nation's economy, and the world's economy, are flows of money, and money is just a record—information about people's assets. Your position in line at the local theater, similarly, is information about how long you've been standing there and when your turn will come. "So it's not true that there is just the computer, and we say, 'Gee, physics has reappeared up there.'" Rather, physics reappears all over; just about *every* level of existence is information. That's what the universe is made of, after all.

This answer assures me of one thing, at least: it's time to leave Fredkin's island; I've heard all this before.

The rain is coming, sweeping across the ocean as a solid sheet of fine roughness. It is a hundred yards away, fifty, ten, right on top of us, banging down on the corrugated steel overhead. Ed is exultant. He peers out eastward and spots a source of even greater gratification. "There's some heavier rain coming across," he says. He does some mental calculating and adds confidently, "It will be here in about two minutes."

Joyce walks up to the table. She and Richard have been swimming over at Honeymoon Beach, which is hidden in a beautiful little cove on the island's west side, and they barely made it back before the rain began. Ed is happy that Joyce got Richard to swim and asks how he did. He did fine.

Suddenly there's no more pounding. Ed looks out at the ocean with irritation. "The main rain went over to Virgin Gorda, dammit. That's only a couple of hundred gallons."

On the brighter side, now I won't miss my flight.

Ed, as always, reads the pilot's checklist, confirming that all the up switches are up and the down switches down. "Okay, passenger briefing." He points behind my seat without looking up. "Back pocket has your life jacket." He looks at the ground around the plane. "Switch on!" he yells, just in case anyone is standing unseen near the propeller. "Clear!" The engine starts, the propeller turns. He taxies the plane down a concrete ramp no bigger than a two-car driveway and into the water, then accelerates against the ocean's grain. After pounding across a couple of dozen waves, we are airborne. Still stalking the Great Quote, I try to elicit a retrospective appraisal of his life. It's generally negative: "I don't think I've done as well as I could have, certainly. I should have gone the regular route and gotten my Ph.D. and worked like a demon."

Should I point out that, by his own analysis, if he had gone the regular route, he wouldn't know that the universe is a computer? No. What's the point of arguing with him? He's just like his father. Instead, I toss out my standard lull-in-the-conversation query. "I have one more question," I yell over the engine noise. "What is the meaning of life?" He doesn't miss a beat. "It has to do with intelligence and information and all that," he yells back. "I think our mission is to create artificial intelligence. It's the next step in evolution." I write this down on a three-by-five index card, to be filed for future reference.

PART II

INFORMATION

IN FORMATION

CHAPTER
NINE

WHAT IS INFORMATION?

Forty years ago, if a person said "damn," you could safely infer that something had gone really wrong. The word was reserved for things like stubbed toes, missed trains, and the discovery that a spouse had somehow accumulated matchbooks from several dozen local motels. Then people started using it for lesser occasions: running out of milk, missing the first five mintues of *Perry Mason*. As *damn* was thus diluted, its original function was assumed by *shit*. *Damn* became the equivalent of *darn*, which then faded out of the picture altogether. By the end of the 1970s, though, *shit*, too, had fallen prey to overuse. In its place came various crude references to sex, and eventually even these expletives became commonplace. The search for alternatives is well under way and has yielded great bursts of creativity. But meanwhile, in mainstream America, far from the cutting edge of profanity, the crisis grows: it is getting harder and harder to vent deep anguish.

At work here is a general principle: any word or phrase used too loosely loses meaning. Thus, *wonderful*, *fantastic*, and *awesome*—which at one time had specific and separate meanings—now mean merely "very good."

The weakening of words through overextension is a particular danger in academia. Scholars, by their nature partial to powerful concepts and thus to broadly applicable terms, sometimes end up spreading them invisibly thin. Among the words now on the endangered-terminology list are *information* and a handful of allied words like *message*.

Even in popular parlance these terms have run the risk of excessive application for some time now. At the turn of the century they already referred to everything from classified ads to love letters, and they have since expanded to encompass the eleven o'clock news and junk mail. But more striking than the growing number of technological things that everyone puts under these labels is the growing number of *biological* things that scientists put there. This century, as human societies have spent more and more time making information, and making things that make information, more and more scientific emphasis has been placed on the information that makes people, and the information that *is* people—and, indeed, the information that makes, and is, all living things.

In 1901 epinephrine (also known as adrenaline) was isolated, and there followed the incremental discovery of how richly it and other hormones supplement the nervous system as conveyers of regulatory "messages," and of how discriminately cells "interpret" such messages. Then, in the 1950s, the structure of deoxyribonucleic acid was discerned. As its mechanisms of replication came to light, biologists began talking about the "transcription" of DNA into RNA (ribonucleic acid) and the "translation" (via the genetic "code") of "messenger" RNA into amino acids. The proteins that those amino acids constitute were termed the "expression"—even the "meaning"—of the genetic "instructions." This invasion of biology by the language of language appears to be far from finished. A pharmacologist at UCLA recently proposed the creation of a new discipline: "pharmacolinguistics." Niels K. Jerne, upon accepting the 1984 Nobel Prize in Medicine, delivered a lecture called "The Generative Grammar of the Immune System." Ed Fredkin hasn't convinced many people that the subatomic world consists of information, but his belief that information pervades all *organic* levels of organization appears to be catching on.

The possibility that information is a fundamental biological concept is exciting. It has long been an important concept in economics, sociology, and social psychology, and it would be nice to see the natural and social sciences speaking the same language for a change. But before getting too excited, we should make sure that *information* hasn't gone the way of *damn* and *awesome*—lost all meaning through loose use. What, after all, do DNA, insulin, and neurotransmitters have to do with tax forms, telegrams, and the utterance "partly cloudy with a chance of rain"—or, for that matter, with each other? What *one thing* does the word *information* mean?

This question is not to be trifled with. For all its superficial simplicity, it is deeply messy. In fact, just to get to a point where we can mull it over without making fools of ourselves, we have to return briefly to the realm of physics for a crash course in the second law of thermodynamics and its conceptual offspring, entropy. But that is all right, because there are other reasons to become conversant in entropy. One is that such conversancy is chic. Many people in California and Colorado who used to sit around talking about the striking parallels between quantum mechanics and Eastern philosophy now talk about the striking paradox posed by life's stubborn persistence in the face of the second law's nihilistic sweep. The second law is even a fashionable literary motif. It plays a leading thematic role in Thomas Pynchon's novel *The Crying of Lot 49*, and has been picked up by enough other writers to make Pynchon something of a guru within the genre. In fact, his reputation as an entropy expert was approaching mammoth proportions when, in 1984, he admitted that he really didn't know much about it. "Do not underestimate the shallowness of my understanding," he warned in a preface to an anthology that included "Entropy," a short story he had written in 1959. "Since I wrote this story I have kept trying to understand entropy, but my grasp becomes less sure the more I read."

There's your second law: slippery. But it is worth trying to grasp—not just because Pynchon and all those people in California and Colorado are trying to grasp it, but because they are exercising sound judgment in doing so. The second law truly warrants the effort of apprehension. It is at once a straightforward and a subtle idea, depressing and uplifting. In it lies a good portion of what science has to say about the meaning of life.

In one of its several incarnations, the second law states that, generally, structure decomposes: dirt clods disintegrate; clouds of gas dissipate; hot spots and cold spots fade into one lukewarm blob. The amount of entropy in the universe—the randomness, the disorder—never decreases, and just about every time anything *happens*, it increases.

To make matters worse, the amount of usable energy declines as entropy grows. The reason is that energy, for people to get a grip on it, has to reside in structure; a nice, distinct, hot body, such as the core of a nuclear reactor or the base of a car's cylinder, is useful because the heat can be harnessed as it spontaneously spreads up and out. Once all the spreading out is done, there is nothing left to be harnessed. The energy's structure has disappeared, a victim of the second law. To be sure, the energy still exists; the heat that was once so neatly concentrated is now spread thinly, more or less randomly, over surrounding space. But this thinness renders it useless; heat you can't get ahold of is wasted heat. So one way of looking at the second law of thermodynamics is as another good-news/bad-news joke, except in reverse. The bad news is that in a jillion or so years, after the second law has taken its entire toll, the universe will have no form whatsoever: no stars, no planets, no mountains, no trees, no us—just a featureless expanse of low-grade energy. The good news is that we'll be better off dead anyway, since there won't be any way to keep our TVs running.

One common misconception about the second law is that evolution violates it. Life, the flawed argument begins, not only preserves structure but multiplies it, giving rise to ever more elaborately ordered organisms, from bacteria through earthworms all the way up to people, who in turn produce castles, cathedrals, and shopping malls; since the second law generally erodes structured things, it must, in the realm of life, be suffering at least a temporary setback. The problem with this logic is that the second law applies only to isolated, "closed" systems. The system of life is open; it receives energy from external sources (the sun, bacon cheeseburgers, and other low-entropy things) and is free to expel high-entropy waste products into other open systems, such as sewers. While it is true that a growing dog or cat or person, by developing coherently structured organs, is increasing the amount of order in the immediate vicinity of its spinal column, this gain is more than outweighed by the disorder the organism discharges into its environment. And eventually that disorder will catch up with us, assuming the universe isn't plugged into some infinite power supply and doesn't drain into an ever-expanding cosmic dumpster. It appears, in short, that we are all unwitting accomplices of the second law of thermodynamics. Like people sinking in quicksand, we doom ourselves more surely as we struggle more vigorously.

But look at the bright side. Even if living things don't violate the second law, they do represent token resistance to it. By coexisting with a law that is generally opposed to coherence, they embody, at least, a bit of irony, which is more than can be said for most other collections of molecules. If we don't defy the letter of the second law, we certainly defy its spirit.

But this is getting ahead of the story.

By the turn of the twentieth century, scientists had a fairly solid understanding of entropy. They knew it was disorder, they knew it lacked useful energy, and they knew it was gaining on them. They even had a precise mathematical definition of it, one that endures to this day, couched in symbols that evoke vague but distinctly uncomfortable memories of trigonometry and calculus. Then something strange happened: thermodynamics intersected with the study of information.

In 1948, Claude Shannon, an engineer at Bell Laboratories, published a paper called "The Mathematical Theory of Communication." Shannon's aims were practical. He wanted to know, for example, how engineers can encode information so that it resists erosion by the white noise encountered in telephone lines. Such analysis called for the formulation of general laws of information transmission, and this, in turn, called for a good working definition of information. So Shannon invented one. The odd thing about his definition was that, when translated into mathematical symbols, it bore a striking resemblance to the definition of entropy. In fact, the two were identical.

Interpretations of this fact differed, and they still do. Some people think it a remarkable coincidence with uncanny overtones; to disappear beyond the horizon in search of a definition of information and then stumble onto a formula from already charted terrain seems almost like voyaging to a virgin planet and finding a Burger King. But really, the "coincidence" isn't all that astounding. With even a general understanding of how Shannon defined information, and of how physicists had defined entropy, the necessity of the convergence becomes clear.

Shannon's definition said, in essence: the more uncertainty there is about the contents of a message that is about to be received, the more information the message contains. Shannon is not talking here about the *meaning* of the information—he astutely avoided this treacherous subject—but about the symbols in which the meaning is encoded. If you're receiving Morse code, the first dot or dash to come over the wire carries a trifling amount of information, since you were certain it would be one of the two. (Assuming the two are equally likely, the one that does appear carries one "binary unit," or "bit," of information.) But if you're watching an Associated Press news story arrive at a teletype machine, the first letter carries much more information, since it could have been any of twenty-six symbols, and you were commensurately uncertain about its identity. This idea of quantifying information in proportion to uncertainty seems to have perfectly reasonable implications. It takes more than one dot or dash per letter to represent the alphabet in Morse code, and naturally a letter contains more information than any one of its constituents.

Entropy, as defined by the end of the nineteenth century, also centered on the idea of uncertainty. In a highly ordered, low-entropy system—such as, say, a glass of pure water—there is very little uncertainty as to what any tiny little region would look like if immensely magnified; there would be lots of H_2O molecules floating around. But a system higher in entropy, like a mud pie, does not admit to such certainty; the knowledge that it is a mud pie doesn't enable you to confidently guess the identity of any one molecule, because mud consists of lots of kinds of molecules, all mixed up.

Given the centrality of uncertainty in the definitions of information and entropy, the mathematical resemblance between the two should not be surprising. If you still find it even moderately puzzling, imagine that you are on a game show called *Name That Molecule*. You are standing there, in front of a home audience of millions, staring at a mud pie. To win the Winnebago, you have to guess what kind of molecule is going to be randomly selected from the mud pie by a blond woman with a large and persistent smile. Her announcement of the molecule's identity is, in Shannon's terminology, the message you are waiting for, and the fact that you are so uncertain about it means that it has a high information content. This fact also means that the mud pie is fairly high in entropy. If you had to guess which molecule would be selected from a glass of pure water, a system much lower in entropy, you would be certain of winning the Winnebago, and could await her announcement calmly; it would not contain any new information. The lower the entropy, the less information there is.

Actually, you can look at the whole thing in exactly the opposite way, too. You can say that the low-entropy, highly ordered system has a lot of information—or, at least, a lot of *conspicuous* information; just by looking at the glass of water you can "read" every message that could originate from it. In other words, there is lots of microscopic information ("H_2O molecule, H_2O molecule, H_2O molecule . . .") embedded in its macroscopic description ("glass of pure water"). The mud pie, on the other hand, carries little conspicuous information; no sequence of molecules can reliably be deduced from the description "mud pie."

So, really, information can be equated with entropy or with the negative of entropy. It is a matter of personal preference. Each convention makes sense in its own way, and so long as scientists remember which one they're using, either will work. The important point is that, regardless of whether a negative or positive sign is attached to the string of symbols representing information, it is identical to the string representing entropy, because both are quantifications of uncertainty.

Shannon himself sometimes spoke of information *as* entropy, but Norbert Wiener, who independently formulated the same definition of information at roughly the same time, lent his weight to the alternative convention. In the introduction to his book *Cybernetics*, he wrote: "Just as the amount of information in a system is a measure of its degree of organization, so the entropy of a system is a measure of its degree of disorganization; and the one is simply the negative of the other." This convention has since gained ground, particularly among laymen. Popular books and magazine articles on information theory, cybernetics, and related areas of cosmic speculation commonly equate information with order and, by loose inference, with any sort of structure or form.

One notable thing about this equation is how far it is from being new. The first definition of *inform* in the *Oxford English Dictionary* is "to give form to, put into form or shape." The *OED*'s earliest example of the word's use in that sense comes from 1590, when Edmund Spenser wrote in *The Faerie Queene* about "infinite shapes of creatures . . . informed in the mud on which the Sunne hath shynd." Only metaphorically did *information* come to have anything to do with communication; to "inform" a mind or a belief or a decision was to impose form on it, bring ordered knowledge to it. Thus, Joseph Butler, a minister, could write in 1736 of "our reason and affections, which God has given us for the information of our judgment and the conduct of our lives." And Thomas Jefferson could write in 1813 of having read a book "with extreme satisfaction and information."

Notwithstanding its etymological resonance, the equation of information with form does not quite meet our present needs; though well suited to the purposes of Claude Shannon and Norbert Wiener, it has shortcomings when construed as a definition of the kind of information we're interested in. Granted, it is true that the things we call information do have form. The patterns of ink in newsprint and the sound waves in "Hello" are well defined. The molecules of hormones, too, are distinctly structured. Even smoke signals have more form than garden-variety smoke. But these facts have little to do with Shannon's technical definition of information; they merely reflect the fact that anything serving as a signal *must* have structure of some sort, since it has to stand out from its background in order to be perceived—whether by a person or a cell or some other living thing. (Shannon did address the necessity of structure in signals, but not in the same train of thought that led to his definition of information.)

Besides, even if—for whatever reason—everything we call information does have form, it is hardly true that all form qualifies as information. Take a brand-new tire, for example. It has form, and it has, in its macroscopic description ("new tire") lots of microscopic information ("rubber, rubber, and more rubber"). Still, is a tire information? There are people who would say so with a straight face. There are professors of physics who would pat the tire warmly and say, "That's a lot of information you're looking at there."

These people are not of much help to us. For their purpose, it may make sense to equate information with order. But their purpose, really, is not so much to analyze information as to analyze order; the kind of information they talk about in pursuit of this purpose— "physicist's information," it might be called—is not the kind we're interested in. We're interested in "real-life information"—information used by organisms (including us), whether internally or externally.

Another way of making the same point is to say that the kind of information physicists talk about—the kind found in glasses of pure water and new tires—doesn't *function* as information in the conventional sense of the term (except, maybe, when physicists are sizing it up). What we're looking for is a definition of information that isolates those things which, in the course of their daily business, *do* function as information. A tire is not functioning as information when it is rolling down I-95, but a newspaper is functioning as information when it is being read. DNA and hormones, similarly, routinely serve as information—or, at least, so we are told by the people who assure us that information is indeed what they are. If we can find a definition of information that encompasses DNA and hormones, and newspapers and radio waves, and doesn't encompass a lot of junk, like tires, we will have reason to believe these people. For now, all we can say is that form is a necessary but not nearly sufficient criterion for real-life information.

As it turns out, even the search for a more discerning definition of real-life information involves entropy and the second law of thermodynamics. In fact, it involves the most famous example of the second law around: Maxwell's Demon.

Maxwell's Demon is the brainchild of James Clerk Maxwell, the Scottish physicist who articulated the unified electromagnetic theory, which showed electricity and magnetism to be different faces of a single force. The demon appeared in a letter from Maxwell to fellow physicist Peter G. Tait in December of 1867 and four years later in Maxwell's book, *Theory of Heat*. Maxwell imagined the existence of two adjacent vessels—something like mouth-to-mouth mayonnaise jars—separated by a partition with a sizable hole in it. The second law of thermodynamics, Maxwell noted, states that if one of these two vessels is filled with hot gas and the other with cool gas, their temperatures will converge; through random motion, energetic, fast-moving molecules on the hot side will dart across to the cold side, while languid, slower molecules meander from the cold to the hot side, until the average molecular velocity is the same on both sides of the boundary and entropy is at a maximum. Once in this state of thermal equilibrium, the gas cannot—or, at least, is staggeringly unlikely to—divide itself into a hot half and a cold half. That, after all, would amount to the spontaneous emergence of order from chaos and of usable energy (in the form of a distinct, hot region) from useless energy. Both of these the second law frowns on.

But suppose, Maxwell wrote, that a "demon" were inside the apparatus, manning the passage between vessels A and B. And suppose he had sharp vision and very quick reflexes. When a slow molecule approached the hole, he could cover it by closing a tiny door with idealized, frictionless hinges. When a fast molecule approached, he could uncover it. Maxwell wrote: "He will then, without expenditure of work, raise the temperature of B and lower that of A, in contradiction to the second law of thermodynamics."

It is tempting to forgo careful thought about Maxwell's idea and opt instead to make fun of it. After all, there *are* no demons. And even if there were, any such creature, however tiny, would, like all other forms of life, suffocate or starve to death if left long enough inside a closed chamber. Upon the demon's death, entropy would start to grow again, and the system would return to thermal equilibrium. And, for all we would know, the temporary decline in entropy would have been due somehow to the import of order and energy in the form of the demon. In short, sticking a little man into the experiment is cheating.

To be sure, Maxwell could respond to such criticism by asserting that the demon, being a demon, was no ordinary creature and could survive without oxygen and food. And you can't argue with him there. Obviously, anything endowed with divine powers could violate the second law of thermodynamics. And if cows had wings . . .

Needless to say, this kind of ridicule is unfair. Maxwell was, like all of us, a prisoner of the times. His thought was hemmed in by the science and technology of his day. Back then, it was impossible to describe with any precision what was going on inside the demon's head. At its most concrete, thinking was thought of as something somehow physical that happens inside the black box known as the brain. At its least concrete, it was considered an immaterial process, an exercise of pure mind, immune to the laws of physics. Information, the stuff of thought, seemed similarly ethereal; it was something that passed invisibly from the outside of the black box to the inside, where it underwent some sort of convolutions.

In the decade after Maxwell wrote about the demon, the telephone would be patented, and in the following century the computer would come to life. Increasingly, scientists would see information acquisition, processing, and transmission as fundamentally physical processes, involving the motion of electrons, sound waves, and the like—and thus involving matter and energy. The demon would then be ready for careful scrutiny. And he would not fare well.

In essence, what the demon is doing is acquiring information about which molecules are fast and which are slow, and storing it by permanently separating the two. He takes microscopic information—which comes to him in the form of photons, the particles of light that bounce off the molecules and into his eyes—and turns it into macroscopic information: two different vessels, one labeled "hot, fast molecules" and the other labeled "cold, slow molecules." Maxwell had implicitly assumed that this valuable information—information that permits a system to swim against the tide of entropy—could come for free, with no expenditure of energy. But twentieth-century analysis revealed that it could not; somewhere in the course of all the shuffling around of information the demon has to pay a price.

Opinions differ on where, exactly, the price has to be paid. Some people say that at least part of it must be paid very early. Since the system is closed (in accordance with the requirements of the second law), light cannot enter from outside, they note; the two mayonnaise jars are painted black, and the demon, just to *see* the molecules, will have to bring his own flashlight, batteries included. As the flashlight dissipates energy, it will increase the entropy, even as the demon is fighting furiously to do exactly the opposite.

This analysis has been widely accepted for a long time, but now a growing number of knowledgeable people consider it mistaken. In recent years, careful thinking about the physics of computation—the same line of thought, in fact, that led Fredkin and Charles Bennett to their reversible computers—has led them to the conclusion that it is not, strictly speaking, in *acquiring* information about each molecule that the demon uses energy. Rather, it is in *erasing* the memory of each molecule to make room in his brain for the next observation; as all reversible-computation buffs know, it is in the *loss* of information that energy gets dissipated and entropy created. And if the demon tries to dodge this fact by bringing in a huge brain with lots of memory space, then that space, originally in pristine, highly ordered condition, like a blank slate, will be gradually messed up by the memories registered on it. This growth of cerebral entropy, Bennett has written, outweighs any shrinkage of entropy elsewhere in the vessel.

This revision is of largely academic interest. It doesn't change the moral of the story as conventionally formulated: the demon cannot, all told, reduce entropy. And it doesn't change the more specific moral of the story that we are interested in: to create even the *illusion* of entropy reduction—to pile up lots of order in one place while sweeping disorder under the rug—the demon must process information. In the end, there is no escape from the second law, and it takes information even to buy time.

This moral suggests a second stab at a definition of real-life information. Apparently, information not only *has* structure; it is a prerequisite for the creation of structure—and for its preservation. It doesn't merely *embody* order; it advances order and maintains it. Information lies not just in form; information lies in formation. It is the stuff that leads the fight against the spirit of the second law.

This definition has some intuitive appeal. Lots of things commonly called information do help create or preserve order. DNA certainly seems to qualify, and hormones help keep organisms functioning in an orderly manner. At the social level, too, order is imposed by all sorts of information: speed limit signs, movie listings, a drill sergeant's barking. And these examples are not as metaphorical as they may sound. Populations of cars and moviegoers and soldiers, like populations of molecules, can be analyzed quantitatively; we can figure out how much information about their "microscopic" states can be inferred from a simple "macroscopic" description. Thus, in knowing that a battalion of soldiers is neatly arrayed on the parade grounds, we are calculably more certain about each individual soldier's location than we are in knowing that they are milling around somewhere on the base. A more subtle, but nonetheless genuine and theoretically quantifiable, kind of order lies in the smooth flow of cars that traffic signals maintain. Movie listings, similarly, lead to the orderly assembly of people at a prearranged place and time.

All of this notwithstanding, some things we call information don't lead to order. Try, for example, yelling "Fire!" during a movie. Or try doubling the numbers on all speed limit signs and changing all stop signs to yield signs. Apparently, the stuff we commonly call information can lead to order but can also lead to chaos.

So these two aspects of real-life information—its "form" and its "formative power"—do not quite amount to a *definition* of it, though they seem to capture much of its significance. Perhaps we could say information is something that has form and is a prerequisite for form; not *all* information leads to form, but the creation of all form involves information.

Unfortunately, this won't work either. The sad fact is that form is sometimes created by things very few of us would call information. The sun, which has form, extracts moisture from diverse damp nooks and crannies and assembles it into large, fairly homogeneous clouds, thus creating form, but few people, aside from Ed Fredkin, would call evaporation an informational process. Similarly, a snowflake, by collecting and neatly arranging ice crystals during its descent, continually supplements its own form, but I do not like the idea of calling a snowflake information.

No, information of the sort we're interested in—real-life information—*isn't* a prerequisite for the creation of form. All we can confidently say about real-life information so far is that it has form and is sometimes involved in the creation of form. Not a very impressive definition. In the end, we may have to settle for something no more rigorous than the definition of pornography laid down by the late Supreme Court Justice Potter Stewart: "I know it when I see it."

But even if a fully satisfactory definition is out of the question, aren't there any additional attributes of information that might at least shed some light on it? Aren't there any things found in the information we use every day—information such as "partly cloudy with a chance of rain"—that might be found in hormones and DNA? One candidate leaps immediately to mind: meaning.

There is cause for caution here. Many gifted philosophers have spent careers thinking about meaning, and the collective results suggest that it is a more elusive concept than information by a few orders of magnitude; the contemplation of it is much less likely to bear fruit and much more likely to leave the contemplator sitting alone in the corner of an off-campus café, mumbling unintelligibly. If we are smart, we will doggedly resist any impulse to think closely about meaning.

On the other hand, if we were smart, we probably would not have gotten bogged down in the contemplation of information in the first place.

TEN

WHAT IS MEANING?

If you follow your family tree down through recorded history and beyond, through the iron and stone ages, through the eras of *Homo habilis* and *Australopithecus afarensis*, down to the primates-at-large branch, then along that branch to the thicker, mammalian branch, then down the mammalian branch to the tree's trunk, then down to the base of the trunk (passing offshoots of reptiles, amphibians, and fishes along the way), you will find the bacterium—mother and father of us all.

This we can say with a fair degree of confidence. But when we try to look beyond the bacterium, to the very bottom of the family tree, things get murky. It isn't clear what sorts of configurations of matter were the early precursors of bacteria. Some scientists think that in the beginning a rudimentary strand of DNA, having formed haphazardly, haphazardly began making copies of itself. Some think the first DNA was a descendant of RNA, which later, through an unknown injustice, came to play second fiddle. Some think both DNA and RNA were preceded by clay crystals that, with serendipitous assistance from other elements, got their pattern preserved in a form that later served as a template, which then forged identically arrayed crystals, which then repeated the cycle. All that seems certain is that at the very base of the tree of life is form. In a sea of shapelessness, something with structure appeared and began getting its pattern copied.

By assuming, for the sake of illustration, that this structured thing was primitive DNA, we can speculate about how the copying initially worked. A precellular strand of DNA would have resembled a string of beads floating around in a sea of unstrung beads. Each bead on the string had a natural affinity for unstrung beads of complementary structure, and thus eventually found itself paired with a partner bead that fit it snugly. Once these complementary beads were all in place, they constituted a string of their own, owing to a similar structural affinity among themselves. The whole apparatus now had the look of a zipped-up zipper: two vertical, parallel chains bound by horizontal links. The vertical bonds proved the stronger of the two; after being jostled around for a while in the primordial sea, the whole thing came unzipped. Both the original and complementary strings of beads were now free to begin construction anew—producing, respectively, a copy of the complementary string and a copy of the original string.

Even in this crude form, DNA seems to satisfy, at least vaguely, the measly criteria for information we have been playing around with: it has form and can give form; it is a highly nonrandom arrangement of matter, and it creates other such arrangements. Like Maxwell's Demon, it does this by sifting through little chunks of matter and choosing the ones that meet its specifications. To be sure, the word *choosing* is a bit metaphorical. What the genetic material actually does is fairly passive: it just sits there while ill-suited beads pass it by and well-suited beads latch on to it; the laws of chemistry do all the work. (In this sense, human DNA, too, is ultimately passive; our construction depends on just such passing chemical attractions and is thus the work not so much of genes as of their environment and universal laws. This fact becomes stranger the longer you think about it.) But however unassertive the primordial DNA, if it hadn't been there, no additional form would have gotten created. It is fair to say, then, that even the simplest strand of DNA weaves coherence out of randomness.

As it turns out, neither the form embodied in DNA nor the form arising from it is the immediate reason that people commonly refer to DNA as information. This habit has more to do with a couple of rough analogies between DNA—modern DNA, that is, the kind that builds organisms—and more familiar forms of information.

The first analogy is with architectural blueprints. Since genes, like blueprints, lead to a major construction project, why not call genes the same thing we call blueprints—information? Both seem to function as instructions, after all, and everyone agrees that instructions are a kind of information.

The second analogy—between the beads on the DNA and the letters of the alphabet—resides at a level of finer detail. Along a modern strand of DNA, there are four kinds of beads. In real life, of course, these are not beads but chemical bases: adenine, cytosine, guanine, and thymine. But, since the word *base* doesn't conjure up a very vivid image (at least, not a very appropriate one), it is easier to think of them as beads, beads of four different colors—say, *a*uburn, *c*hartreuse, *g*old, and *t*urquoise. It is after being copied onto a strand of RNA—which, aside from some conspicuous but inconsequential details, is then identical to the strand of DNA—that these beads begin to powerfully suggest a comparison with the alphabet. Three beads in a row—a "triplet"—will be "translated" into an amino acid. The ribosome, the cellular machine that does the translating, will, for example, upon "reading" a triplet consisting of *g*old, *c*hartreuse, and *a*uburn beads (in that order), select the amino acid alanine from among the various amino acids circulating nearby. It is as if the ribosome had a decoder, like a decoder for Morse code. Only, instead of saying things like "$- - \cdot\,\cdot$ = Z," the decoder for the genetic code says things like "GCA = alanine." After selecting the alanine, the ribosome moves along the strand of RNA to the next triplet, selects the amino acid it calls for, and hooks that amino acid up to the alanine. The ribosome continues in this fashion until it encounters one of the several triplets that serve as "punctuation," at which point the chain of amino acids is complete. The next amino acid will be the first link in a new chain.

These chains form the protein molecules that are the building blocks of life; they constitute much of the matter that constitutes us, and enzymes, a special class of protein molecules, regulate all kinds of chemical reactions that have to take place if we are to make a respectable showing in our struggle with the spirit of the second law.

It is easy to see why the four kinds of bases in DNA are sometimes referred to as the letters of the genetic alphabet. Like letters, they are distinctive units that have no importance in isolation but become significant when clustered together in a particular order. Indeed, the word *significant* has been taken quite literally in some corners. A few distinguished thinkers have referred to the amino acids, the proteins, and the organism they add up to as the "meaning" of the DNA.

This way of looking at things has borne fruit. Especially useful has been a distinction between the DNA's "explicit" meaning (the sequences of amino acids that its various segments will yield) and its "implicit" meaning (the organism that will ultimately emerge from those amino acids). Through discussion of the "context" required for the realization of implicit meaning, the nature of ontogeny has been illuminated. In particular, it has been shown that the common analogy between DNA and a blueprint is in fact hugely misleading. Each step in the process by which genes give rise to organisms is so dependent on the environment, and on the results of similarly contingent past steps, that even the best informed embryologist could not sketch the contours of an unknown organism after "reading" its genetic text; though the DNA's explicit meaning is written all over it, the only way to grasp its implicit meaning is to watch things unfold.

Thus, for certain pedagogical purposes, it makes sense to call an organism the meaning of its DNA, or to say, analogously, that the meaning of a bare-bones, primordial strand of DNA is the complementary bare-bones, primordial strand it creates. Still, for our present purpose, it is of no help to use the word *meaning* in this almost metaphorical way. Our present purpose is to isolate the birth of meaning in a more conventional sense of the term—to find out where in the tree of life symbols began to *function* meaningfully, the way words function in everyday conversation. My guess is that, contrary to natural expectation, this spot is very near the base of the tree, not all that far from the primordial strand of DNA that does nothing more than make copies of itself. Finding that spot may help us put a finer point on our inchoate definition of information. But to find it we must figure out what we mean by meaning.

"A dead dog on the kitchen table." If someone asked me what that string of symbols means, I would, without trying to be flip, say, "It means that there is a dead dog on the kitchen table." If pressed for a less circular explanation, I might take a more practical tack: these symbols mean that if you walk into the kitchen you will see a table, and on it you will find something with four legs and fur, and if you say "Fetch," it will not comply. This sort of pragmatic restatement of a sentence may sound like a fairly seat-of-the-pants, commonsensical, even naive, approach to the question of meaning, but not so long ago it stood on the frontiers of Western philosophy, part of a revolutionary school of thought called, fittingly enough, pragmatism.

Pragmatism was launched in the late nineteenth century by Charles S. Peirce (pronounced "purse"), an American scientist and philosopher who today is undeservedly obscure. It was then popularized by William James (who, along with several other philosophers, proceeded to warp it beyond Peirce's recognition). In analyzing the function of thought, both Peirce and James took what might be called a Darwinian point of view. The purpose of our beliefs about the world, they said, is to foster rules of behavior that help us survive. As James put it, thought helps us achieve "satisfactory relations with our surroundings." It is not surprising, then, that both men considered the import of our thoughts—the *meaning* of our beliefs—to be closely connected with the behaviors they lead to. The exact nature of that connection is debatable, both because the two men headed off in different directions and because Peirce himself seems to have been of two minds about the subject.

Take the statement "Diamonds are hard." At one point in his article "How to Make Our Ideas Clear," which appeared in *Popular Science Monthly* in 1878, Peirce says this sentence means that diamonds "will not be scratched by many other substances." This suggests that the meaning of an assertion lies in a prediction about the consequences of certain specifiable behaviors: "Diamonds are hard" means that if you rub a diamond against lots of other substances, few of them will scratch it; "rocks are heavier than air" means that dropped rocks will drop; "the bridge over the alligator pit can support no more than fifty pounds" means that if you walk across the bridge it will be your last walk. Thus construed, pragmatism is essentially an affirmation of the scientific method: any meaningful declarative sentence can be recast as a prediction about the results of one or more experiments. Peirce here appears to be saying simply that thinking like a scientist is a good way to clarify the meaning of ideas—even ideas not traditionally considered scientific.

At times, though, his point seems to be slightly but significantly different: to say that diamonds are hard means not just that *if someone does* scrape a diamond against other substances it will not be scratched; it also means that, in recognition of this fact, people will not try to scratch diamonds with glass or gold or steel—and that, moreover, people who want to scratch other substances (glass cutters, for instance) will sometimes use diamonds. Peirce wrote, "What a thing means is simply what habits it involves."

So which is it? Is the meaning of a message the behavior it induces? Or is the meaning of a message a prediction about the consequences of experiments? Maybe these two ideas can be reconciled. Consider the statement "Apples do not contain cyanide." The behavioral meaning of this message—the "habit it involves" is the continued consumption of apples by people. But whether the statement is *true* depends on what happens next: whether the people eating the apples—and thus conducting a de facto test of the statement—die of cyanide poisoning. You might, in a gross but instructive oversimplification, say that the *meaning* of a message is the behavior it induces, and the *truth* of a message depends on whether that behavior has consequences beneficial to the behaver.

In one of the more bizarre twists in the history of philosophy, William James took this shorthand, hybrid version of pragmatism literally and carried it into the realm of religion. In essence, James said: If you believe in God, and this belief brings solace and optimism and improves the quality of your life, then it is true; God exists. (James seems to have been untroubled by the fact that, since a single belief can have different effects on different people, his logic implies the simultaneous truth of the statements "God exists" and "God doesn't exist.") This born-again pragmatism—if it feels good, believe it—embodied only the most extreme of many distortions pragmatism had suffered since Peirce originally used the term. To distance himself from the confusion, Peirce added two letters to the word, renaming his own philosophy "pragmaticism." The new word was ugly enough, he hoped, to discourage the kind of unsolicited adoption that had corrupted the old one. (This decision apparently did not offend James. He continued his various efforts to assist Peirce professionally and personally, including occasional subsidy during Peirce's final, penurious years.)

With sufficient intellectual contortion, pragmatism could perhaps be used to reinforce the conception of meaning I dismissed near the beginning of this chapter—the idea that the meaning of a strand of primordial DNA is the complementary strand of DNA it creates. After all, if the meaning of a string of symbols is whatever activity they lead to—as pragmatism, loosely speaking, maintains—and a strand of DNA leads to the assembly of complementary DNA, then doesn't this complementary DNA, or, at least the assembly of it, then constitute the DNA's meaning? In a word, no. There is something that primordial DNA doesn't have that "Diamonds are hard" does have—namely, diamonds: something out there in the real world that the symbols represent.

Of course, you could argue that the DNA *does* "represent" something; it represents the strand of complementary DNA it is about to create, just as a blueprint represents an unbuilt building. This is a permissible thing to say, I guess, but it amounts to saying that what the DNA *represents* is the same thing it *leads to*. In contrast, the kinds of symbols that Peirce and James had in mind ("signs" is what Peirce actually called them) lead to one thing (an individual's behavior) by virtue of representing something else, something out in the environment (apples, diamonds, etc.). Indeed, the whole function of these symbols is to serve as intermediaries between individual and environment—to *inform* the individual's behavior with news about likely consequences.

Along this link between individual and environment lies the most satisfactory reconciliation of Peirce's two different senses of the word *meaning*. A red traffic light is a symbol that means, in one sense, that if you venture heedlessly past it, you stand a reasonable chance of getting killed, because cars are likely to be zipping back and forth beneath it. This meaning is a statement of fact, a representation of the environment. The red light's second meaning—that you will, in fact, stop—grows directly out of the first meaning; it is the behavior that brings the individual's goals (such as staying alive) into harmony with the environment, as represented. Similarly, the phrase "partly cloudy with a chance of rain" means, first, that if you go outside without an umbrella, you may well get wet and, second, that, in recognition of this fact, you are more likely to carry an umbrella than you would be if blue skies were forecast.

It is time to admit that the immediate outcome of this digression into the history of philosophy is anticlimactically simple: bare-bones DNA, the sort of stuff that is at the very base of the family tree, somewhere between life and plain old matter, doesn't have *pragmatic* meaning. It may in some sense represent something, and it definitely leads to something, but those two things are either identical or almost identical (depending on how you look at it); the DNA is not involved in tailoring the latter to the former in the way that traffic lights and weather forecasts are.

If primordial DNA lacked pragmatic meaning, then when in natural history did such meaning happen on the scene? As soon as DNA started doing anything very complicated, such as building a moderately sophisticated little cell in which to house itself.

We don't know exactly what sort of bacteria our ancestors were, but we presume they were comparable in some important respects to the sort still floating around—such as *Escherichia coli*, the most copiously studied bacterium, if not organism, in the world.

To build and maintain this bacterium (or, for that matter, *any* ordered system) requires energy. Energy often comes in the form of carbon, and it is no coincidence that *E. coli* typically inhabits places rich in sources of carbon, such as human intestines. But sometimes, through bad luck, *E. coli* finds itself in an environment with little or no carbon. Help is on the way—or, rather, the bacterium is on the way to help.

The absence of carbon prompts the bacterium to synthesize a molecule called cyclic AMP and secrete it internally. (This "prompting" mechanism is actually not so much a trigger as a suppressor. The enzyme that makes cyclic AMP does so automatically *unless* inhibited, and the presence of carbon-containing molecules, like sugars, inhibits it.) The cyclic AMP molecule, once built, attaches itself to a protein, and the resulting complex then attaches itself to a segment of the bacterial DNA. This binding alters the expression of the DNA, inducing the synthesis of various proteins appropriate to an energy-poor environment. In particular, the DNA produces a protein essential to the construction of flagella, the propulsive tails that bacteria use on occasion as an alternative to aimless flotation. After the flagellum buds, it begins whipping around and transports the bacterium to another environment—one that, with luck, will be well stocked with carbon.

E. coli and cyclic AMP occupied center stage in an article called "The Metabolic Code," which appeared in *Science* magazine in 1975.

The author, Gordon Tomkins, a biochemist who had died that year, contended that the cyclic AMP molecule is a "symbol," one that "represents a unique state of the environment." You could even go a step further than Tomkins did, and say that the molecule possesses meaning. After all, the situation has all three prerequisites for pragmatic meaning: a symbol (the cyclic AMP molecule); something in the environment that the symbol represents (a shortage of carbon); and something alive that, upon processing the symbol, behaves so as to reconcile its well-being with that environmental condition (by heading elsewhere). Though the bacterium may not be acutely conscious of the logic behind its actions, its behavior is nonetheless parallel to that of someone who reaches for an umbrella upon hearing a weather report. (Strictly speaking, the comparison is not between the cyclic AMP molecule and the weather report but between the cyclic AMP molecule and the neural impulses that bring the weather report to the brain. After all, the cyclic AMP originates within the organism and travels only from its periphery to its headquarters, the DNA. For now, the imprecision in the analogy is not important; the main point is that all of these forms of information—words, neural impulses, and cyclic AMP—can be said to have meaning because they represent states of the environment and induce behaviors appropriate to those states. As we will see in chapter 17, cyclic AMP has in some species grown more analogous to a weather report as evolution has proceeded.)

If the cyclic AMP molecule can indeed be said to have meaning, then we have an interesting situation here. On the one hand we have something that is a precursor of life but doesn't seem to quite qualify as life itself (the no-frills, precellular version of DNA, nothing but a template for a template), and it does not appear to process meaningful information. On the other hand, we have something that *is* commonly considered a form of life (a bacterium), and it *does* process meaningful information. The obvious temptation is to conclude, without any of the critical scrutiny that so often spoils great generalizations, that *meaning* is the hallmark of life.

This idea dovetails nicely with some traditional thinking about the essence of life. In circles where people ponder these issues, it is sometimes said that the hallmark of life is purposeful behavior; all living things act as if they had a well-defined purpose—to replicate the pattern in their DNA and to achieve various goals subordinate to that mission, like finding food and mates. And, if you think about it, you will see that pursuing any purpose entails processing meaningful information.

But before reaching that conclusion, you must first think about the word *purpose* carefully, lest you embarrass yourself with its loose use. Attributing purpose to things is a tricky business at best, and doing it incautiously has produced a lot of ideas that are now considered silly. For example, it was long an article of faith—and still is in some societies—that the purpose of rain is to make crops grow. The logic behind this belief is simple: after the rain falls, the crops do grow. In modern societies, however, the conventional wisdom is that the rain was falling long before plants grew on this planet and that, though plants have since come along and put it to good use, the rain has no inherent *purpose*. We have become suspicious of attributing purpose to things just because certain other things typically follow them. This smacks of teleology; it sounds as if we are viewing causality as a pulling, rather than a pushing force, as if we are saying that tomorrow's crop growth in some sense *causes* today's rainfall.

Indeed, fear of teleology has even made some scholars skittish about imputing purpose to people. Thus, if you say something perfectly reasonable like, "My purpose in going to McDonald's was to eat Chicken McNuggets," you could get upbraided by a graduate student in philosophy: "Are you saying that the eating of the Chicken McNuggets *caused* you to go to McDonald's? Are you saying that future events determine present ones?" You might reply that your going to McDonald's was caused not by your eating but by your wanting to eat, which preceded the trip. But then you would catch flack from the behaviorists, psychologists who insist that psychology, if it is to be a science, must make no reference to internal, subjective states such as wanting.

At the dawn of World War II, behaviorism was ascendant—all other attempts to make psychology "scientific" having failed in one sense or another—and fear of teleology was widespread; the concept of "purpose" was not in vogue. But war brought a desperate need for new information technology, and new information technology suggested a way to fit the concept of purpose into a scientific outlook. One of the engineers of this reconciliation was Norbert Wiener, the man who would later define information in terms once reserved for entropy just as Claude Shannon was doing the same thing. Wiener was a prodigious thinker (Ph.D. from Harvard at age eighteen, etc.) and he looked like one: he wore thick glasses, and his clumsiness struck fear and pathos into the hearts of lab instructors. His eyesight had kept him out of the army in World War I, and he made his contribution to the Allied effort during World War II at his desk, figuring out how to increase the accuracy of antiaircraft fire through automation. He thus came to contemplate the idea of *feedback*—and, in particular, negative feedback.

Negative feedback is information about the discrepancy between the present state of affairs and the state of affairs a system aims to bring about; it is called "feedback" because it is fed back into the system to bring the system nearer to this goal. A good example of feedback in action is a kiss. If you are going to lean over and kiss someone on the lips—especially someone you have never kissed before—you would be well advised to keep an eye on that person's lips as you move yours toward them. That way you can continuously absorb feedback—visual information about where this pair of lips is in relation to yours. The molecule of cyclic AMP in *E. coli* is another example of feedback; both it and the information about the relative location of lips reflect a discrepancy between the ideal situation (an abundance of carbon and four adjacent lips, respectively) and the present situation (no carbon, no lip contact), and in both cases this information effects change that reduces the discrepancy. For examples of feedback closer to home, consult your thermostat or toilet. The level of the float in the toilet tank represents the amount of water in the bowl. So long as there is a discrepancy between this amount and the ideal amount, the incoming flow of water continues; when the discrepancy is zero, the information reflecting the zero (the level of the float, that is) shuts off the flow.

Wiener popularized the idea of feedback in *Cybernetics*, published in 1948, but back in 1943, in a paper titled "Behavior, Purpose, and Teleology," he and two coauthors had put the idea in a context of particular relevance to the present discussion. One point of the paper was that, with the aid of the concept of feedback, purposeful behavior could be explained in a concrete, scientific manner, without attributing present events to future events, and without reference to states of mind. Thus, a working toilet is a perfectly mechanical thing, and it complies entirely with the laws of physics. We presume that it has no sensation of "wanting" to be filled and that there is no sense in which the future state of fullness is "causing" the alteration of its present emptiness. Rather, the information is set up to flow in such a way that the toilet will behave *as if* these things were the case. So too with the bacterium: the causes and effects lying behind its relocation have nothing mystical about them; perfectly concrete information, abiding by the laws of physics, just flows in a way that fosters the illusion of guidance. And, really, it's not an illusion; these flows of information *amount* to guidance.

So too with humans: though they may have "desires" and "plans" and "conscious goals," we needn't talk about these messy things in order to explain their purposeful behavior. We can explain it—in principle, at least—by reference to a completely tangible, if confusingly elaborate, flow of information: photons, sound waves, neural impulses, and the like, interacting in perfect accordance with the laws of physics. The information flow in humans, like that in bacteria and toilets, is simply *set up* to yield behavior that is purposeful, whether or not you want to call it that.

Of course, the words *set up* suggest that we're just begging the question: Who set these systems up? It sounds as if, when we get to the bottom of this thing, we will find an immensely purposeful being, a king-size example of teleology. In fact, though, what we will find is not a being but a process; organic evolution set the bacterium up and set us up, and we then proceeded to set the toilet up. A simple redundancy—that what survives survives—has, in conjunction with the tendency of DNA to impel occasionally imperfect copies of itself through life cycle after life cycle, built immensely complex machines and endowed them with purpose. And now those machines are inventing their own machines and endowing *them* with purpose. Natural selection, clearly, is a process that deserves our respect.

Wiener noted, in his 1943 paper, the intimate association between purpose and information. He declared that *any system*—a person, a thermostat—must, in order to flexibly pursue a goal, employ negative feedback. Like Shannon, Wiener generally avoided the subject of meaning, but if he were here today, and were forced to read about Charles Peirce and the pragmatic conception of meaning, he might well agree that negative feedback is necessarily meaningful information. It represents a discrepancy between a goal and the present situation, and thus makes a statement about the way things are: "the level of carbon is inadequate"; "the amount of water in the toilet is less than ideal"; "the distance between my lips and those lips is too great." While we feel certain that neither a toilet nor a bacterium consciously understands the messages it receives, it is clear that they behave as if they did; the meaning of the messages is imparted to them in a pragmatic sense. So if purpose is indeed the threshold between near-life (like unadorned, primordial DNA) and real-life (like bacteria), then so is the processing of meaningful information. So next time someone asks you if your life has any meaning, any purpose, you can emphatically and truthfully answer yes to both questions, even if you are at that very moment perched on the edge of the existential abyss. It will be an inside joke.

Having failed, in the last chapter, to come up with a good definition of information, we can now scale down our ambitions and try to come up with a good definition—or at least a serviceable ad hoc definition—of *meaningful* information. Here goes. Meaningful information is that which has form, can help create or maintain form, and does so by representing states of the environment and inducing behaviors appropriate to them.

I am not wild about this definition. Upon close inspection, it would be seen to encompass lots of things that I would rather it didn't. Still, it is sufficiently interesting to warrant exploration. In particular, we might dredge up the question that originally inspired our search for a definition of information and slightly rephrase it: Could the term *"meaningful* information" be applied to the various things that get called information these days—everything from weather forecasts to hormones to human DNA?

Weather forecasts certainly have form—sound waves or newsprint or patterns on a TV screen. They also create and maintain form. By keeping people from dying of frostbite, they maintain the form of human bodies; and, by sending scores of people searching for their umbrellas simultaneously and permitting them to arrive at work on time in spite of inclement weather, they lend form to the patterns of daily social behavior. The forecasts achieve this, obviously, by representing states of the environment (future states, actually) and by inducing behaviors appropriate to them. So weather forecasts do indeed seem to qualify as meaningful information.

Other kinds of information exchanged by humans, though, do not comply so neatly with this definition of meaningful information. Tax returns, shopping receipts, and record albums fit it so loosely that it would take a doctoral dissertation to justify their inclusion. And some utterances—lies, for example—obviously fall short of the accurate representation that is part of the definition.

We will take up this issue of dishonest communication in chapter 17. And there, too, it will become apparent that hormones fit very nicely into our ad hoc definition of meaningful information. But what of DNA? We have already tried, and failed, to find meaning in barebones, primordial DNA. Does modern-day DNA—the kind that builds plants and people—qualify as meaningful information?

It certainly has form and gives form, and it is strikingly reminiscent of the alphabet, to boot. Still, like primordial DNA, it appears to lack the thing that gives cyclic AMP and weather forecasts the right to be called meaningful. It doesn't represent states of the environment (at least, not in any straightforward way).

If DNA isn't meaningful information, perhaps there is another kind of information it is—namely, a program. Like a cellular automaton, or any other computer program, it embodies rules—algorithms—for converting input into output. When stated in elemental form, these rules sound none too momentous: if segment X of the DNA is transcribed and then translated, series Y of amino acids will be strung together into protein Z; if segment A is activated instead, protein C will result. But when viewed in a broader sweep, these rules take on significance; many little, seemingly trivial rules can amount to a single, larger, clearly consequential rule: if there are no sources of carbon in the environment, a flagellum will bud.

So even if DNA is not itself meaningful, it does *process* meaningful information. The meaningful cyclic AMP molecule conveys its representation of the environment to the bacterial DNA, and the DNA takes appropriate action. Or, looked at another way, the DNA takes reports ("There is a shortage of carbon sources") and turns them into instructions ("Build a flagellum"). In similar fashion, the DNA in the cells that are us processes information that represents *their* immediate environments. That is the only way that bone cells, hair cells, and skin cells—all of which are born with the same general-purpose DNA, after all—"know" which identity to adopt; their DNA receives reports about what sort of neighborhood they are in, and it then instructs them to act accordingly. The existence of these three kinds of information—reports, instructions, and programs—is, no doubt, one of the things that has been making it so hard for us to arrive at a neat and simple definition of information.

Perhaps the best way to characterize the relationship between DNA and meaning is to say that DNA is the *source* of meaning. It takes information about the environment and turns it into behavior— thus realizing meaning in the pragmatic sense of the word. DNA is the place where the two sides of meaning meet, the place where reports become instructions. DNA is thus what first gave meaning to life; or, perhaps, what first created meaning, and therefore life; or what first created life, and therefore meaning. In any event, it is very impressive stuff.

PART III

EDWARD O. WILSON

CHAPTER

ELEVEN

UNITY

Lecture Hall C, one of five large bowls in Harvard's Science Center, is mostly filled. More than two hundred students have settled into their orange-red, cloth-upholstered, theater-style seats, unzipped or unsnapped their down coats, and loosened or removed their winter scarves. The fold-out desktops are in use, supporting spiral notebooks, elbows, Malto Milks cookies (in Handy Pax), half-pints of whole milk, coffee in styrofoam cups. Course B-15—Evolutionary Biology—is almost history: four more lectures to go, counting this one, and then comes Christmas break—a would-be holiday sabotaged by the certainty that final exams will follow. Up front—so far from the back row that I might as well watch the lecture on TV (which students have the option of doing later; a videotape camera stands in the right-hand aisle)—is a blackboard. On the blackboard is one word, about six feet long and a foot high: *sociobiology*.

The man standing in front of the blackboard is attached to that word almost as firmly as Einstein is to *relativity*, Bell to *telephone*. There is little he could do, short of assassinating a president, that would keep it out of the first paragraph of his obituary. Not that he would *want* to break his bond with it. Since the publication of his five-pound, 697-page book, *Sociobiology: The New Synthesis*, in 1975, he has devoted untold hours to the word. He has used it liberally in radio and television interviews, written articles and books that elaborate on its meaning, and donated blurbs to other books that contain it, all the while trying to fend off the nasty connotations his detractors would like to affix; no, he has insisted, sociobiology is not racist or sexist; no, he is not a fascist or a social Darwinist.

The natural inclination is to take his word on this; E. O. Wilson looks anything but evil. At age fifty-six, he still has an adolescent gangliness about him. He walks long-leggedly and gestures long-

113

armedly, and his chest doesn't quite fill out his sport coat; the lapels bow outward at times, and his necktie does the same above the tie clasp. Wilson's face, like the rest of him, is small-boned, and he has long, dark eyebrows that jump up when he stresses a point. His glasses, circa-1959, have little silver diamonds in the corners of the frames, like the pair worn by Robert Carradine in his starring role as Lewis in *Revenge of the Nerds*.

Today I am looking for bits of Billy Graham in E. O. Wilson. Wilson is known for the enthusiasm with which he writes and speaks about sociobiology, and there has been speculation that this zeal has something to do with his Southern Baptist upbringing. In fact, back in 1981, when I first interviewed him, he volunteered the observation that Sunday sermons may have taught him, and many southern writers, "what it takes to reach an audience." Now, years later, after hours of additional interviews in his office and over the phone, I am finally getting around to writing about him, and it has occurred to me that sitting in on one of these lectures might provide a serviceable literary device: I could describe him thrusting his hands this way and that, pounding the lectern, preaching furiously about sociobiology. Then I could pepper the rest of the profile with religious imagery, thereby giving it a unifying theme.

Unfortunately, Wilson turns out to have a fairly sober style of lecturing, and I find myself straining to salvage my motif. Isn't there something vaguely ministerial in the way he holds his hands outstretched, palms up, to frame rhetorical questions? Is that fervor I see in the fists he clenches for emphasis? And what about the eternal pacing, back and forth, as he laces the two sides of the lecture hall equally with eye contact? Sort of like an evangelist roaming the rostrum—right? There's some truth in all this, but not quite enough for my purposes. Whatever rhetorical techniques Wilson picked up on Sunday mornings in the South have been tempered by more than three decades in academia. The most dramatic thing that can honestly be said of his style is that it evinces a flair for showmanship.

And it is a good show he puts on. On a screen above the blackboard appears a series of illustrative and sometimes arresting images: graphs, charts, human beings, hamadryas baboons, wild dogs of Tanzania. Sometimes he points to these with a flashlight that emits a narrow and intense beam of red light, and sometimes he uses his hand, sweeping it outward and upward like Carol Merrill revealing what's behind door number one. He employs the provocative overstatement to hold attention, the well-timed quip for comic relief; the phrase "a giant step for apedom," in the right context, draws hearty laughter. Throughout, he displays an eloquence rare among scientists; his sentences, though sometimes distended, are consistently populated with well-turned phrases. (*On Human Nature*, the sequel to *Sociobiology*, won the Pulitzer Prize for general nonfiction.) E. O. Wilson makes evolutionary biology about as entertaining as it can be made without compromise of content, and the applause at the end of this lecture will not be perfunctory.

Actually, when I leave aside Wilson's gestures and focus on his words, I can find something that *borders* on fervor. It is a self-certainty with a caustic edge, even shades of militance. In arguing that moral and ethical intuitions are shaped by the genes, he satirizes the alternative view—that morality is an "angelic code," "divinely given" or discerned by humans with their "superior intelligence." And in explaining why many scholars in the social sciences and humanities resist the use of ideas from evolutionary biology, he wastes no time on diplomacy. These people, he says, have a "disjunctive, dissociative" way of looking at life, a way that doesn't bring order to our understanding of it. No wonder the social sciences are indulging in an "orgy of self-flagellation" (faint laughter from the audience) over their failure to meet the standards for theoretical power long taken for granted in the natural sciences. No wonder "an increasing number of social theorists and philosophers" are reaching the conclusion that they have met the enemy and it is them—that they are ignoring at their peril the biological basis of human behavior and failing "to look to the next level of organization . . . for those unifying lawful generalizations that are needed before you can produce a true theory."

The next level of organization. More than most scientists, E. O. Wilson thinks about science. He thinks about the rules of conduct, the drive to discover, the social organization of the enterprise and its conceptual organization. And there is one image, implicit in his thinking, that almost single-handedly accounts for his work over the past dozen years, with all its ups and downs—the Pulitzer Prize, the

National Medal of Science, the alliances formed and the friendships broken, the praise that he relishes and the criticism that has hurt him. This is the image of a solid structure rising certainly into the air—a tower, or a skyscraper, or, perhaps, a pyramid. At the top of the pyramid are the social sciences. Below them are the biological sciences. Below them is chemistry, and below it is physics.

It is not a new metaphor, by any means; it has occurred to thousands of thinkers over the years. But Wilson takes it more seriously than most. As science progresses, he believes, each level of inquiry will be seen to rest on the level beneath it in a fairly literal sense: its laws will follow from the laws below, almost as surely as the Pythagorean theorem follows from the basic assumptions of plane geometry.

This belief is known as reductionism. At the lower levels of organization, reductionism will not get you into hot water. Many of the laws of chemistry have been reduced in a rigorous way to the laws of physics, so reductionism is not, among chemists, an arguable philosophy. (That is not to say they are enthusiastic about it. Because of this reduction, some of the work they once did with test tubes can now be done by mathematicians with a knowledge of physics.)

It is only in the higher regions of the pyramid of knowledge that reductionism becomes disputable. Does biology literally rest on chemistry? Could the behavior of, say, a kidney be predicted with much precision from a knowledge of the molecules involved? How about a brain? Could laws describing a dog's or a chimpanzee's or a person's behavior be deduced from the laws of organic chemistry?

These questions, in addition to being difficult, are loaded with philosophical consequence. At these higher levels of organization, reductionism is allied with determinism, which holds that free will is a myth, that the rest of human history will unfold as inevitably as a cellular automaton, however powerful the illusion that we are directing it with our various "choices." Our inability to predict this predetermined future—or even to predict one person's behavior on a day-to-day basis—reflects, according to determinists, only incomplete data and our ignorance of the principles involved. Humans are very complicated, after all, their behavior guided by millions of microscopic signals per split second. In practice, the action is simply impossible to keep track of. Nonetheless, if each signal is, as we presume, the inevitable result of some other signal or signals, which in turn were grounded with equal firmness in *their* antecedents, then we could, in principle, project the path of behavioral causality forward—somewhat as we could, in principle, trace the path of one of Ed Fredkin's billiard balls backward—and show that even the most ethereal and inspiring of human accomplishments is the product of mechanical necessity. Thus say the determinists, and thus say the reductionists. And thus are they not the life of many parties.

Further besmirching the reputation of reductionists at the pinnacle of the pyramid of knowledge is the fact that social scientists, like chemists, are not eager to surrender turf. And the less eager they are, the more inclined they are to see imperialistic designs in ideas that drift up from below. Thus, within the social sciences, the word *reductionism* has taken on distinctly pejorative, and offensive, connotations: a reductionist is someone who offers simplistic biological explanations of human behavior, often with hostile intent. E. O. Wilson, though once insensitive to the mores of the social sciences, has become intimately and painfully acquainted with these connotations since 1975.

Whether Wilson really is a reductionist in this sense of the word is still a subject of debate. But he is without doubt a reductionist in the pure sense of the word—and proud of it. Since 1981, in fact, he has been pushing a theory that is just about as reductionist in spirit as a theory can be. It is an attempt to encompass in one fell swoop most of the pyramid of science, an attempt to formulate laws linking molecular biology to psychology, anthropology, and sociology. The theory comes complete with a mathematical model designed to track the effects of changes in the gene pool on human culture and, conversely, of cultural change on the gene pool. With a large and arcane system of equations, Wilson is trying to cover the sizable gap between the information that shapes us and the information we shape.

Needless to say, this theory—the "gene-culture theory," set forth in a book called *Genes, Mind, and Culture*, which Wilson coauthored with a young physicist named Charles Lumsden—has not found favor with truckloads of social scientists. Many biologists are also skeptical of it. Further, and more significantly, a number of *socio*biologists are unhappy with the theory, a fact that has done nothing to shore up Wilson's standing as leader of his crusade. In fact, it is not inconceivable that the gene-culture theory will turn out to be the only thing other than assassinating a president that could keep Wilson's word out of the first paragraph of his obituary.

The theory has shortcomings, Wilson concedes. But it is an article of faith for him—faith in the old-fashioned, Southern Baptist sense of the word—that eventually it will be vindicated. "I believe in a kind of unity," he says. "A unity of knowledge."

CHAPTER

TWELVE

ANTS

Edward Osborne Wilson was born on June 10, 1929, in Birmingham, Alabama, a town that, like many others, he would leave shortly after entering. His father was an accountant, first for the Rural Electrification Administration and then for the U.S. Army, and both jobs entailed frequent relocation. By the time Ed got his high school diploma, at age seventeen, he had attended sixteen schools. Almost all were in southern towns: Mobile, Brewton, Decatur, Evergreen, Pensacola—never farther north than Washington, D.C.

Wilson declines to say much about his father or his mother, and he does so consistently. "I don't want to cut too deeply into that," he said in 1981 when asked what his father was like. Five years later, when asked if he wanted to say anything about either parent, he replied, "I don't want to cut to that depth." He will disclose only that his parents were divorced when he was seven and that, for financial reasons, he lived with his father and stepmother—although for a time during high school, while they were away, he lived with a family friend.

Aside from an odd inability to memorize poetry, prose, and song (even today "The Star-Spangled Banner" comes only with great difficulty), Wilson was endowed with solid intellectual equipment. And, though he did not especially like schoolwork, from an early age he had—for reasons still unknown to him—a strong drive to excel. He skipped the third grade, an honor that had the unintended effect of reinforcing the already powerful logic behind solitude: in addition to being an only child and changing schools more often than grades, he was always, as he puts it, the runt in the class.

He found companionship and continuity in natural history. The towns and faces changed, but there were always *National Geographic*s to read, and woods to roam, and organisms to observe. Wilson has a theory about unsettling childhoods being conducive to careers in biology. He cites the experience of two German entomologists he knows well. "Both were children in the Second World War, their families torn up and partly killed. In both cases the kid was left on his own resources out in the woods and developed this deep affinity for natural history." Wilson doesn't want the comparison with his life to be drawn too strictly. "Obviously, I was coddled compared to them. But there might be certain parallels."

It was in boyhood that Wilson first experienced "the naturalist's trance," as he called it in his partly autobiographical book *Biophilia*. Alone in the woods, free from the demands of human society, he could enter a world oblivious to his. He still can. "I need a bigger fix now," he says. "When I was eighteen I could get all excited just getting out into a swamp in Alabama. Now it takes a rain forest in Brazil. But I still get the same rush, the same emotional high—hoping to find new things, exploring."

Occasionally Ed managed to turn a schoolmate into a "part-time zoologist." One convert was Ellis MacLeod, now a professor of entomology at the University of Illinois. In the fifth grade, MacLeod and Wilson lived blocks apart in Washington, D.C. On weekends they rode the bus to the Smithsonian Natural History Museum or walked to the National Zoo or to Rock Creek Park. At the park, wielding butterfly nets made of broomsticks, coat hangers, and cheesecloth, they bagged red admirals, fritillaries, and mourning cloaks, which would later be killed, pressed, and immortalized as representatives of their species.

Ellis remembers Ed writing stories whose heroes were animals and reading them before the entire student body. In spite of literary renown, though, Ed was shy. "I don't recall that he had many close friends, if any, and I didn't either," says MacLeod. "I think we were each other's close friend, and virtually our only friend." Laughing, he adds, "We were both a couple of creeps by contemporary standards—you know, those crazy sorts of people who go around picking up insects. I know Ed, in sixth grade recess, once just absolutely blew every kid's mind by letting a wolf spider walk over his hand." By junior high, when MacLeod visited the Wilsons in Alabama, Ed had moved up from spiders to "beetles, big showy things . . . praying mantises." A year or two later he assembled a backyard collection of snakes. They lived in chicken-wire cages and dined on the fish and tree frogs that Wilson dutifully procured. Inquiring neighbors were treated to tours of the collection. "I was rather locally famous for my devotion to snakes," Wilson recalls.

But his fate was not to study snakes. One day, at age seven, the year of his parents' divorce, he was fishing from a wharf and he yanked one of his catches out of the water with such uncontrolled force that its fin entered his right eye. The result was a traumatic cataract that left the eye nearly useless, permanently dulling his perception of depth. But the vision in his left eye remained acute, especially at close range, a fact that steered him toward the study of the small-scale. "I am the last to spot a hawk sitting in a tree," he has written, "but I can examine the hairs and contours of an insect's body without the aid of a magnifying glass."

Ants first caught Wilson's attention at Rock Creek Park. The memory is still clear: upon picking apart a rotting tree stump, he beheld a dense colony of *Acanthomyops*. Their golden bodies glittered in the sun, and the air was pungent with their characteristic smell, citronella. The next year, after moving to Alabama, he embarked on a systematic study of ants, collecting specimens of *Odontomachus insularis* and *Iridomyrmex humilis*. His career aspirations were beginning to take shape: he would work for the U.S. Department of Agriculture's extension service—drive around in one of those government-green pickup trucks and advise farmers about insect enemies and allies. At age thirteen Wilson made "my first publishable observation" (though he would not publish it for years)—that the Mobile area was rife with a fire ant once confined to South America: *Solenopsis invicta*.

"On several counts ants can be regarded as the premier social insects," Wilson wrote in *The Insect Societies*, the book that, upon publication in 1971, established him as one of the world's foremost entomologists. In elaboration, he speaks with awe about the ability of these simple little creatures to constitute societies of such complexity, size, and relentless efficiency.

Consider *Eciton burchelli*. These army ants possess a discipline, a sense of mission (to judge by appearance, at least), and, collectively, a cohesion, that few human troops attain. At night they bivouac;

hundreds of thousands of ants link their bodies together, forming a solid cylinder several feet in diameter, with the queen and her brood secure in the center. At daybreak this body dissolves into temporary chaos, out of which emerges a marching column. The smaller, more agile worker ants take the lead, while bulkier ants, the soldiers, follow along on the shoulder of the trail. Leaves rustle underfoot, and grasshoppers make popping sounds as they dart into tree trunks in self-defense—often to no avail; many of those that succeed in escaping the ants are eaten by ant thrushes, scavenger birds that follow the offensive assiduously from a series of nearby perches.

If you let your eyes fall out of focus, says Wilson, the ants look like a single body, ten or fifteen yards long and more than a yard wide. It grows like a tree, extending and then dividing into two, three, five branches, which subdivide and merge until the ants form a united front, an arc perhaps twenty yards across, that moves methodically forward. This band seeks and destroys tarantulas, scorpions, beetles, roaches, grasshoppers, alien ants—even insufficiently agile snakes and lizards. Death comes through stinging or asphyxiation. Dismemberment and transport proceed without deliberation; half a dozen workers will carry a scorpion's tail back fifteen yards to the "booty cache," from which, by day's end, it will be taken to the bivouac—traveling, all told, the length of a football field.

Less terrifying but no less impressive than the army ants is *Atta sexdens*, a species of agricultural ant. An *Atta* colony employs an intricate division of labor to simultaneously plant, tend, and harvest its staple crop, fungus. The largest workers forage and return to the nest with leaves, which smaller ants cut into manageable sizes. Still smaller ants chew these fragments into little pellets and then turn them over to another caste, whose members embed them in the earth. Upon these beds members of yet another caste implant tiny tufts of fungus. Tending the crop is left to the smallest ants of all; they lick it clean, weed out foreign fungi, and extract strands periodically to feed their sisters.

Although an *Atta* colony doesn't resemble an organism quite so strikingly as does a colony of army ants, it does possess some of the properties of organisms: division of labor, the selfless devotion of the parts to the whole, and their utter dependence on it. In fact, the same could be said of all ant colonies. William Morton Wheeler—

Harvard's great turn-of-the-century entomologist, whose heir E. O. Wilson is—contended that the ant colony *is* a kind of organism, a "superorganism." He wrote: "Like the cell or the person, it behaves as a unitary whole, maintaining its identity in space, resisting dissolution." It met his definition of an organism as "neither a thing nor a concept, but a continual flux or process"—"a complex, definitely coordinated and therefore individualized system of activities, which are primarily directed to obtaining and assimilating substances from an environment, to producing other similar systems, known as offspring, and to protecting the system itself and usually also its offspring from disturbances emanating from the environment." Thus the queen, in Wheeler's view, was merely a colonial egg, albeit "a winged and possibly conscious egg."

This blurring of the line between society and organism is a delicate matter, and it lies behind some of the criticism to which E. O. Wilson has been exposed. There is something unsettling—to Western sensibilities, especially—about a society that works with such mechanical precision and blind individual sacrifice. Fascist Italy springs immediately to mind, as do Communist China and even modern Japan, which, for all its cheap and trusty cars and computers, is a society whose efficiency some Americans find eerie. Further, it doesn't take much imagination to see a parallel between *Atta*, with its caste system, and a human society in which people are assigned at birth to a permanent socioeconomic station. Wilson, to be sure, has never held up ant societies as worthy of human emulation, but he has written that they are more nearly "perfect," in a strictly biological sense, than are mammalian societies, and at the left end of the political spectrum—where criticism of Wilson has more than once originated—such language has a reactionary sound to it. (It may, then, seem ironic that Wheeler, who took the superorganism idea very literally and seemed to derive a certain aesthetic pleasure from it, was a socialist. But in a way this makes sense. The nagging problem faced by socialist and communist societies has been that humans are selfish; if they are not recompensed in proportion to their work, they tend not to work very hard. An ant society has no such problem. Its altruism is deeply programmed.)

The ant colony's unity, in addition to being politically suggestive, is baffling, at least on the face of it, because it is realized in such mindless fashion. Ants are capable of only a small array of stereotyped behaviors and are unable to tell one nestmate from another. How does such individual stupidity add up to societal intelligence? In Wheeler's day, this was a profound mystery surrounding all socially complex insects—wasps, termites, and bees, as well as ants—and it provoked some fairly weird speculation. Maurice Maeterlinck, the mystically minded Belgian who won the Nobel Prize in Literature in 1911, attributed the cohesion of bee colonies to something called "the spirit of the hive." This spirit, he wrote in *The Life of the Bee*,

> regulates day by day the number of births, and contrives that these shall strictly accord with the number of flowers that brighten the country-side. It decrees the queen's deposition or warns her that she must depart; it compels her to bring her own rivals into the world, and rear them royally, protecting them from their mother's political hatred. . . . Finally, it is the spirit of the hive that fixes the hour of the great annual sacrifice to the genius of the race: the hour, that is, of the swarm; when we find a whole people, who have attained the topmost pinnacle of prosperity and power, suddenly abandoning to the generation to come their wealth and their palaces, their homes and the fruits of their labour; themselves content to encounter the hardships and perils of a new and distant country.

Wheeler considered Maeterlinck's ideas fair game for derision, and he was not much more charitable toward other immaterial explanations of organic unity, such as "entelechy" (which, according to its inventor, the German philosopher and biologist Hans Adolf Eduard Driesch, was a "whole-making" factor outside of space and time). But even Wheeler, in trying to explain the insect societies' coherence, was forced to consider "psychological agencies like consciousness and will." In the case of ants, the correct and more down-to-earth explanation would not be confidently advanced until a half-century after Wheeler wrote. E. O. Wilson, by then a young professor at Harvard, would find the critical piece of evidence while playing with one of his childhood friends from Alabama, *Solenopsis invicta*.

Natural history was not the only unifying theme in Wilson's boyhood. There was also church. His parents were "sit-down Baptists . . . not the Holy Rollers," but even the more staid Southern Baptist congregations are not very staid, and there was an intensity about the services. He remembers being engrossed in the sermons, at once a worshipper and an analyst, feeling repentant, God-fearing, and spiritually enraptured, yet all the while dissecting the rhetorical techniques that evoked these responses so reliably.

Decades later, in *On Human Nature*, Wilson would argue that an inclination toward religious belief is in our genes. Among the genetically rooted cognitive traits undergirding it, he wrote, are the intuitive dichotomization between the sacred and the profane, intense attention toward charismatic leaders, and a capacity for trancelike states. He paid particular attention to a ritual found the world around: the confirmation of identity, during which an adolescent is "transformed by religious experience," granted membership in "a group claiming great powers," and given "a driving purpose in life compatible with his self-interest."

Wilson went through the Southern Baptist version of that ritual in his mid-teens. Near the end of a service at the First Baptist Church of Pensacola, he accepted the "invitation"; as the congregation sang "Softly and Tenderly Jesus Is Calling," he walked up the aisle and told the minister he was ready to accept Christ as his savior. Immersion came in a few weeks, at a baptismal service.

Barely two years later, the ritual's philosophical foundation was displaced. "It was in college that I came up against it in its full grandeur," he says of the theory of natural selection. "I was completely taken by that as an organizing principle." He was struck not just by the power of the principle but by the legitimacy it gave his career plans. The idea that everything about every animal could be accounted for with "a real, scientific, unifying explanation was totally transforming for me."

I once asked Wilson if the word *epiphany* could be accurately applied to such intellectual experiences. "I think there probably is a similar emotional response," he said. "You know, in the typical epiphany, or conversion, the individual says something along the lines of 'I discovered God. Jesus came into my life.' But the outcome of all this is that the individual sees the unity in the universe. Instead of just this fragmented world in which he's doing selfish acts, he sees a purpose for the universe of which he is a small part. And in a tribalistic manner he now submerges himself into the grand plan, a great plan. And that brings a certain very profound peace. Now in a somewhat related way—but with real differences—I think that discovering something of a unifying idea gives you a sense that you do have the key, not to the universe but to the big chunk of it that matters the most to you. In my case, what mattered the most to me was biological diversity."

At the University of Alabama, Wilson found, in addition to Truth, a young assistant professor and two undergraduates who shared his love for nonhuman organisms. "I felt as though I belonged from the beginning," he says. The four of them pored over Ernst Mayr's book, *Systematics and the Origin of Species*, "as holy scripture," and they spent much time collecting data to fit into its framework. Wilson has written: "Our little band of zealots descended into caves to search for trogolophilic crayfish and beetles, skeined streams and ponds for fish, and drove along the highways on rainy nights looking for migrating tree frogs."

There was one disappointment at the University of Alabama: he didn't make the track team. "I would have liked to have been a distance runner literally and physically," he says. "Books like *The Insect Societies* and *Sociobiology* are intellectual marathon runs."

Wilson earned his bachelor's degree in three years and went to the University of Tennessee for graduate work. It was an interesting time to be studying biology there. A quarter of a century had passed since the Scopes trial, but the state remained, technically speaking, an inhospitable environment for Darwinists; teaching the theory of natural selection in the public schools was still against the law. Wilson remembers delivering a lecture about evolution to an undergraduate class and wondering how the students would react to what had been legislatively defined as heresy. After the lecture he found out. A big, hulking boy—a football player, Wilson surmised—approached the lectern. He had only one question: Will this be on the test?

Maybe Tennessee wasn't the place for E. O. Wilson. While at Alabama, he had struck up a correspondence with a Harvard graduate student in entomology named William Brown, and during Wilson's first, and last, year at Tennessee, Brown encouraged him to come to Harvard and take advantage of the largest ant collection in the world.

After two years of graduate work at Harvard, Wilson was elected a junior fellow in Harvard's Society of Fellows. The fellowship afforded him (in addition to illustrious company) three years of study abroad, and he took full advantage. He clarified the geographic distribution of ants in Cuba, Mexico, Fiji, the New Hebrides, and Sri Lanka, and tidied up their taxonomy. From primitive species in New Caledonia, New Guinea, and Australia, he pieced together the evolution of the army ants. He also discovered new species of ants, which still bear his name: *Strumigenys wilsoni, Rhytidoponera wilsoni.* Wilson happened upon people, too, but they didn't make much of an impression. "I wasn't thinking about human beings then," he says. "I saw a lot of tribes and the like, but I wasn't thinking like an anthropologist."

After returning to Cambridge, Wilson married. He is not much more talkative about his wife than about his parents. He says of Irene only that he "met her socially" in the Boston area and that their marriage has been fulfilling. They have an adopted daughter, who recently graduated from college. Irene stays in the background of his professional life, and few colleagues have met her; invitations to the Wilson home are almost unheard of, and when he must attend a cocktail party, he goes alone. *The Insect Societies* is dedicated to "Irene, who understands."

T̲hough Wilson never took the idea of the "spirit of the hive" seriously, evidence for a better explanation was a long time in coming; as of 1953, the social cement of bees, wasps, and ants had not been found. That year Wilson attended a lecture by Konrad Lorenz, the Austrian zoologist who founded ethology, commonly defined as the study of animal behavior from an evolutionary perspective. (Wilson has defined sociobiology as the study of the biological basis of social behavior in all animals, including humans, and he takes pains to distinguish it from ethology—and from the numerous other fields with which it extensively overlaps.) During his lecture, Lorenz talked about "releasers"—signals, usually visual or auditory, that trigger stereotyped sequences of behavior in birds and other animals. Wilson immediately saw the analogous role that chemical signals might play in the social insects.

Several years later, after his travels, he tested his hunch. He took a specimen of *Solenopsis invicta* and sacrificed it to science. Seeking the source of the odor that emanates from the fire ant's trails, Wilson removed three organs that seemed likely candidates and washed them. (No easy task; each was about the size of a short and inordinately thin piece of thread.) He then crushed each organ into a pulp. Only the Dufour's gland, at the base of the stinger, proved provocative. When its remnants were smeared on a glass pathway to a nest of fire ants, dozens poured out, headed for the spot, and, upon reaching it, milled around as if looking for something to do.

Wilson had found the source of the chemical trail markers that keep fire ants in line as they forage. And he had discovered that Lorenz's releasers are indeed analogous. This chemical—or pheromone, as chemical signals later came to be called—triggers a fairly complex behavioral sequence; it not only keeps ants on the trail but also persuades them to leave the nest and begin the trek in the first place. It is, in Wilson's words, "not just the guidepost, but the entire message." He has written of his reaction to the experiment: "That night I couldn't sleep. I envisioned accounting for the entire social repertory of the ants with a small number of chemical releasers."

He made some progress toward that goal. He isolated the alarm pheromone of the fire ants and harvester ants, which they emit in the face of a threat, sending their compatriots into a combative frenzy. And he found a "necrophoric substance" that accumulates in an ant's body after death, inducing other ants to take the corpse to the graveyard lest it gum up the societal works. Wilson tainted live ants with the substance and watched as they were carted off prematurely, their resistance notwithstanding—an elegant demonstration that ant societies owe a greater debt to the power of pheromones than to the intelligence of ants.

Not all the glory was to be Wilson's. Even as he was discovering the source and effect of trail markers, Martin Lindauer, a German entomologist, was isolating an alarm pheromone in a species of leafcutter ant. And Lindauer went further, identifying the composition of the substance and thus becoming the first scientist to chemically characterize a pheromone.

As for Wilson's dream: it now appears that a fairly small number of signals does indeed account for the complex cohesiveness of ant societies. It is as if each species had a vocabulary of ten or fifteen messages, and everyone took instructions without question. One message means "Follow me." Another: "On guard! A threat to the public good is present." (This was the message in the scent that Wilson had evoked by picking apart the tree stump in Rock Creek Park; the alarm pheromone of *Acanthomyops* is essence of citronella.) Another: "Help me clean my body; it has so many hard-to-reach places." Another: "I'm dead. Get me out of here; I'll only get in the way." There is even evidence that these messages can be combined. When a fire ant is forcibly detained, it emits both alarm and trail pheromones, arousing other ants and pointing them in its direction. In effect: "I'm in trouble and I'm over here."

Wilson's main theoretical work in insect communication came in concert with William Bossert, then a graduate student and now a professor of applied mathematics at Harvard. The two men, in a paper that Wilson describes as "the first general theory of the chemical and physical design features of pheromones," invented several analytical tools (things like "active space" and "Q/K ratios") and showed that the molecules of various pheromones are well suited by size and shape to their function. Wilson, who concedes that his mathematical skills are irremediably mediocre, has repeated this pattern of collaboration more than once; when he wants to make a major theoretical contribution, he matches his expansive knowledge, conceptual boldness, and sheer enthusiasm with someone else's rigor.

The discovery of pheromones came amid general excitement about communication. The ideas of Claude Shannon and, especially, Norbert Wiener had by the 1950s diffused into biology and the social sciences. Since information and, more specifically, feedback are found in every animal and animal society, prospects for bringing a new precision to scholarship at the higher levels of organization seemed bright. Some scholars went so far as to hope that the terminology of Shannon and Wiener would become a vocabulary for interdisciplinary discourse and thus help fuse the sciences into a coherent structure—perhaps even the unshakable pyramid that reductionists dream of. After all, the first prerequisite for the formal reduction of one set of laws to another is that the two be couched in the same terms; a barrel of apples can't be reduced to oranges. So it was tempting to speculate that information would become the basic unit of analysis in all the life sciences and then the medium of their unification. Wilson was not among the more starry-eyed in this respect, but he was quite

taken by the idea that a single measure of information could be applied
to different species and different levels of organization. He even
quantified the information in the waggle dance of the honeybee (by
which foragers, upon returning to the hive, announce the location of
flowers) and compared it with the amount of information in the
pheromones of ants. The two communications technologies, he
found, were about equally efficient.

The decoding of the messages that bind ant colonies did not send
shock waves through the scientific community. Norbert Wiener,
among others, had already established that coordinated behavior of
all kinds involves the transmission of information. Besides, cursory
inspection of an ant colony suggests that its trails have been invisibly
marked; an ant that ventures off the path behaves remarkably like a
hound dog that has lost its quarry's scent. So even before Wilson had
crushed the Dufour's gland and deciphered its contents, the prevailing
suspicion was that ants use some such trail markers and perhaps other
chemical signals as well. Pheromones were in the air.

The confirmation of this suspicion, while explaining the unity of
ant societies in one sense, left its mystery intact in another. True,
the mechanism of orchestration—the medium of communication—
had been found. But why were such mechanisms warranted in the
first place? Why had evolution built unified, cooperative societies?
To put the issue in formal parlance: the proximate cause of insect
integration was clear, but the ultimate cause wasn't.

This mystery is deeper than it sounds. Asking why evolution
integrated ants so thoroughly is not like asking why it gave them
legs or mandibles. Legs and mandibles make immediate evolutionary
sense: obviously, they help an ant survive. Social cooperation is not
so unequivocally advantageous. Consider the starkest form of cooperation: total self-sacrifice—pure, final, and necessarily unrecipro-
cated altruism. Bees disembowel themselves by stinging intruders.
Ants of the species *Camponotus saundersi* defend the homeland by
detonating themselves; they contract the muscles of the abdomen
so intensely that it ruptures and releases a sticky substance, which
turns the ground into flypaper, stymying aggressors. If natural selection really is the survival of the fittest, why would it preserve

behaviors that are so emphatically not conducive to individual survival?

The question applies equally to less dramatic forms of sacrifice.
When an ant shares food, or devotes valuable time to grooming
another ant, it receives no apparent compensation. Granted, grooming and food sharing—and self-detonation—may bestow benefits on
the colony as a whole. But, as we will see shortly, there is reason to
doubt that behaviors very often evolve for "the good of the colony"
or "the good of the group" or (especially) "the good of the species."
So, beneath the question of how such stupid creatures as ants behave
with such collective intelligence lurks a question that is even more
stubborn and subtle: How did such selfish things as genes get roped
into the cooperative behavior that constitutes the intelligence?

The issue is not confined to insects. Prairie dogs endanger themselves by conspicuously barking to warn fellow dogs of an approaching coyote or hawk, and the meerkats of Africa also sound alarms
upon sighting predators. Many other mammals share food and groom
one another, and human beings have been known to jump on hand
grenades to save brothers in arms. William Morton Wheeler saw the
breadth of selflessness clearly back in 1910. Biologists, he wrote,
must not be preoccupied with the "struggle for existence" but rather
must explore "the ability of the organism to temporize and compromise with other organisms, to inhibit certain activities of the aequi-
potential unit in the interests of the unit itself and of other organisms;
in a word, to secure survival through a kind of egoistic altruism."

When Wheeler wrote, and for decades thereafter, some biologists
considered the explanation for altruism to be not all that tricky. In
fact, even today there are people who will dismiss the problem with
a wave of hand, casually remarking that altruism evolved for the
"good of the society." Such remarks should be greeted coolly. One
common sign of fuzzy understanding of the theory of natural selection
is loose talk about the good of groups. Careful thought shows how
difficult it is for an altruistic trait to survive simply on the strength
of its benefits to the society, or species, at large.

The way evolution works, so far as we know, is that a new gene
or new combination of genes arises—through random mutation or
sexual recombination—and somehow improves the chances that an
organism will survive and reproduce. That is, the gene brings an

"adaptive" trait. For example, if prairie dogs were all bald, and froze to death by the scores during winter, and one of them underwent a genetic mutation that afforded it fur, then the new gene would, other things being equal, proliferate. The prairie dog carrying it would be more likely to survive and reproduce than other prairie dogs, as would any sons or daughters carrying it. Slowly, the gene for furriness would spread across the population. Obviously, something like this actually happened during mammalian history.

But the story is not so simple for genes that benefit organisms other than their own. Imagine a prairie dog colony in which selfishness is rampant, and imagine that a gene for altruism arises. Specifically, suppose the gene inclines the prairie dog to do what prairie dogs in fact do—stand up on its hind legs and sound a warning call upon sighting an invader. How long would such a gene last? Roughly as long as it took its host organism to encounter a coyote. The altruist would dutifully stand up, emit its alarm signal, and, having attracted the invader's attention, get slain and fade into the annals of prairie dog history.

This is not to say that a society full of such individuals wouldn't fare well. On the contrary, it might thrive as no society ever had before. But genetic mutations don't generally appear all across the board; they show up in one or two animals and then have to work their way to wider acceptance. The question, then, is whether a single "warning-call gene," or a handful of them, could pervade the colony in the first place. And the answer appears to be no. There is no obvious reason why altruism should ever get off the ground.

So, all told, biologists who put a lot of stock in "group selection" (the "group selectionists") are right about one thing: if two societies within a single species differ in their degree of altruism, the more altruistic one may outperform the other in Darwinian terms. Indeed, the selfish society may perish while the selfless one prospers, thus leaving the gene pool teeming with altruistic instructions. But these biologists sometimes neglect an important point: if such group selection is the only way for genetically based altruism to spread, there is unlikely to be such a situation in the first place.

Fortunately for the theory of natural selection, it turns out that there is a way altruism can get off the ground *without* relying on group selection. It is called "kin selection." The idea behind kin selection is that the gene, not the individual, is the unit of natural selection, and the interests of the gene and the interests of the individual don't always coincide.

If you are a prairie dog (or are just yourself, for that matter) and you have a recently invented gene—synthesized, say, by your great-grandparents—so do roughly half of your siblings and one eighth of your cousins. Now, suppose this gene is a warning-call gene, and suppose some siblings and cousins are in your vicinity when a predator appears. You get up on your hind legs and tip everyone off to the impending danger, in the process tipping the predator off to your location and getting eaten. This may seem like a very valiant thing for you, and your warning-call gene, to do. But, in fact, the "sacrifice" made by your warning-call gene is no such thing. Sure, the gene perishes along with its "altruistic" possessor (you), but meanwhile, in the bodies of a dozen siblings and cousins, two or three or four or five carbon copies of the gene are carried off safely to be transmitted to future generations. Such a gene will do much better on the evolutionary marketplace than a "coward" gene, which would save itself only to see its several replicas plucked from the gene pool. The theory of kin selection is very much in the spirit of Samuel Butler's observation that "a hen is only an egg's way of making another egg." Only it adds this point: "So is that hen's sister."

It is important to be clear on what the theory of kin selection does and doesn't say. It doesn't say that genes can sense copies of themselves in another organism and direct their own organism to behave hospitably toward it. Genes are not clairvoyant, and they are not little puppet masters that govern behavior on a day-to-day basis. Their main influence on behavior comes through their construction of the brain, which thereafter is in charge. The theory of kin selection says simply that natural selection can be expected to permit the proliferation of genes that build brains that lead to behaviors that are likely to help kin. And this expectation has a very simple basis: genes that do so will, under many circumstances, do a better job of ensuring *their own* survival (or, strictly speaking, the survival of copies of themselves), than genes that don't. Anyone who understands this clearly should thrill at the subtlety and power of the theory of kin selection—and, therefore, at the subtlety and power of the theory of natural selection, of which the theory of kin selection is a corollary.

Note that kin selection can foster behavior that, on any given day, may not help kin. If prairie dogs usually live near relatives, an automatic warning call, sounded without checking to see whether brothers and sisters are actually nearby, may make evolutionary sense, since they so often are. (As it turns out, prairie dogs do, apparently, check; they are more likely to use the call to tip off kin than to save just any old dog.) Similarly, if for millions of years *Homo sapiens* have been likely to grow up near their brothers and sisters, evolution may give them genes that dispose them to acquire an attachment to youngsters near whom they are reared. This attachment may then, in modern society, transfer to adopted siblings or next-door neighbors. Thus, critics of sociobiology are mistaken when they contend, as they occasionally do, that altruism directed toward unrelated organisms is never explicable in terms of kin selection.

Before the theory of kin selection came along, biologists used the term *fitness* to refer to a gene's contribution to the survival and reproduction of the organism containing it. The word was thought to exactly capture the evolutionary imperative: the fittest genes would, by definition, flourish. This assumption—that evolution maximizes individual fitness—was built into the mathematical models of population biologists. Hypothetical genes underlying the sex and hunger drives and selective aggression fit nicely into those models, but hypothetical genes underlying altruism didn't.

The theory of kin selection changed things. Now the models of population biology are based on the assumption that evolution maximizes *inclusive fitness*. This term is broad enough to encompass the gene's total contribution to the survival and reproduction of the information encoded in it, regardless of whether the immediate beneficiary is its particular vehicle or a different vehicle containing the same information. The new math can be used to show how food sharing, grooming, and various valorous behaviors could have a genetic basis, and to define the theoretical limits of such altruism. This logic has been summed up in an anecdote about an evolutionary biologist who was asked whether, after all he had learned about the ruthless genetic calculus programmed into people, he would still, as he had once vowed, give his life for his brother. "No," he replied. "Two brothers or eight cousins."

The theory of kin selection, though implicit in the writings of Darwin, was not cast in mathematical form (nor widely appreciated even in verbal form) until 1964. In that year, William D. Hamilton, a colleague of Wilson's, developed it rigorously and used it to solve—or, more accurately, to suggest a very plausible solution to—the puzzle of the social insects' superorganic unity.

There are now more words on the blackboard—not just *sociobiology*, but *siphonophore*, *hermeneutics*, *zooids*, and *kin selection*. Wilson is adding the name *William Hamilton* and recalling the circumstances under which the theory of kin selection took shape. "Now it was a remarkable achievement in the early sixties by a then-young British biologist named Bill Hamilton, who looked at social insects in a wholly new way—and I might add, just for your delectation, a footnote. At that time he and I were probably the leading students, younger students, of social insects. And he looked at it in a way that would have never crossed my mind."

Hamilton, like other entomologists, was perplexed by insects of the order Hymenoptera—wasps, bees, and ants. Why is it, he wondered, that highly integrated social behavior is found in virtually no insect groups outside of Hymenoptera? (Termites are the lone exception.) What do wasps, bees, and ants have in common that would account for their cohesion?

One thing they have in common is an odd approach to reproduction: some of their eggs yield life without fertilization. In fact, what determines the sex of an unborn ant is whether its egg is fertilized. If it is, then it becomes a female; if not—if it receives no genetic input from the male—then it becomes, ironically enough, a male. This means that when one of these males matures and produces sperm, the genetic information is the same in all of his sperm cells; having come from an unfertilized egg, he has only one set of chromosomes on which to draw. The queen, in contrast, having emerged from a fertilized egg, will have (like humans) two sets of chromosomes—twice as much genetic information as was actually needed to construct her. Since each egg she produces will draw randomly on this store, her eggs will differ from one another.

So ant eggs are just like most eggs in the animal kingdom; any two produced by the same mother have about half of their genes in common. But ant sperm are weird; any two sperm cells produced by the same father are identical. The upshot is this: the various females born of the fusion of these sperm cells and these eggs will be very closely related—much more so than ordinary "sisters." Whereas two human sisters have about one half of their genes in common, two ant sisters share, on average, three fourths of their genes. (Actually, these fractions are misleading. People in fact share much more than half of their genes with *any* given person—and, for that matter, with any given chimpanzee. But fairly *novel* genes, genes that are not yet established in the population, do indeed have a 50 percent chance of residing in the sisters of their human carriers and a 75 percent chance of residing in the sisters of their ant carriers. And novel genes, being on the cutting edge of evolution, are the ones we're interested in here.)

In his 1964 paper, Hamilton argued that the large genetic overlap among ants revises the mathematics of altruism. If it makes genetic sense for a prairie dog to die because otherwise four fertile sisters will each face a 51 percent chance of death, he reasoned, the same sacrifice makes sense for an ant with only three similarly threatened fertile sisters. If it makes sense for a monkey to share a little food with her sister monkeys, it makes sense for an ant to share a lot of food. In general, the logic appears to be slanted toward cooperation and self-sacrifice in the order Hymenoptera. The line between self-interest and the interest of a sibling, always a little blurry, is blurred further in ant, wasp, and bee societies.

Wilson says to his students, "Hamilton said, 'Is it possible that this bias, this kin selection, is the driving force behind repeated origination of this type of higher colonial organization in the bees, wasps, and ants?' And he sold me immediately on this idea. And I was one of the early—I was *the* early defender of this view. And I must say that I've had to concede that Hamilton—even though I think I knew more about social insects—Hamilton beat me to it to produce the main idea, the most original, important idea on social insects of this century. And I had to react the way young Huxley, Thomas Henry Huxley, reacted when he read *The Origin of the Species*. Here was Darwin saying, Look, we can explain all these marvelous things by natural selection. . . . And Huxley's comment was the one that I made: How *stupid* of me not to have thought of that. Why

didn't I sit down and *think* for a few minutes"—he clenches his fists—
"instead of running out in the field and, you know, doing all these
things? Well, anyway"—the students laugh at Wilson and thus with
him—"as a consequence, kin selection, and, you know, this basic
approach, has become central in the development of the field of
sociobiology."

CHAPTER
THIRTEEN
PEOPLE

E. O. Wilson likes military imagery. He has described to his students the convergence of biology and the social sciences in these terms: "Now we are approaching a point where biology—the biological imperium, as it were—has come within sight of the parapets and pennants of the social sciences." Once, when asked why his gene-culture theory was cast in such densely mathematical form, he replied, "We had to armor-plate *Genes, Mind, and Culture*." As the book was nearing publication, a small, cardboard, Neanderthal-looking man took his station on Wilson's desk; frozen in the middle of a bold stride, marching as to war, he carried a banner that read GCT FOREVER. This standard-bearer for the gene-culture theory was a gift from Charles Lumsden. It had been modeled after Wilson's "mascot" ant, which was prominently pictured in *Time* magazine's 1977 cover story on sociobiology. The ant—a slightly larger-than-life replica of an ant, actually—still sits in Wilson's office, under a glass dome, supporting a banner that reads ONWARD SOCIOBIOLOGY!

These are not empty symbols. Wilson approaches science like a field general heading into world war. His objectives are large and clearly defined, his strategies are meticulously mapped out, and he pursues them with relentless discipline and a sense that right is on his side. The enemy, ultimately, is not the social sciences per se but the sort of intellectual entropy they represent. "I've always wanted to transform messy subjects into scientifically orderly subjects," he says. "To put things right, so to speak."

Illustrative of Wilson's military acumen is a maneuver he engineered in the early 1960s, one that elevated him from entomologist to evolutionary biologist at large. One purpose of the strategy was to resolve a career crisis and thus pull him out of a prolonged state of mild depression. The source of his depression was reductionism, in the pure sense of the word.

In 1953, James Watson and Francis Crick had become famous by discovering the helical structure of DNA and figuring out, broadly, how it makes copies of itself. Wilson at first found this affirmation of Darwinism inspiring. "It was a thrill," he has written, "to learn that the underlying molecule was in fact quite simple and that straightforward, readily understood chemical principles could be translated upward into an all-but-limitless biological complexity." After a time, though, the double helix turned on him. It drew research funds and young talent into molecular biology, while Wilson's brand of biology received little of either. There was even some question as to whether plain old evolutionary biology still had a reason to live. Why estimate the phylogenetic relationship between two species by comparing bone structure when the exact answer would soon be legible in the genetic text? Why speculate about the role of genes in behavior instead of seeking the answer in the genes themselves? Everything would ultimately be explained from the bottom up, so why bother with the top?

On Harvard's faculty in the late 1950s and early 1960s was James Watson himself. His enthusiasm for molecular biology was comparable to Wilson's present enthusiasm for sociobiology, and his territorial ambitions were even less tempered by tact; he is, one colleague of his told me, a man who almost seems to enjoy making other people suffer. Watson openly argued—in department meetings, among other places —that evolutionary biology had a limited future at best; the real action would come in molecular biology, and that was where the university should put its money. "He just took an extreme view, that's all," says Wilson. "He was very young and rough at the edges. He's famous for being immature and extreme. I hope he's mellowed by this time."

Wilson could feel the hostility of Watson and his "allies" upon passing them in the hall, and he lacked allies of his own. The other evolutionary biologists at Harvard were older men—Ernst Mayr, George Gaylord Simpson—and were in the "consolidation period of their careers," Wilson recalls. He felt alone, stranded between his intellectual forebears and his contemporaries. He found himself wondering whether the molecular biologists weren't right: *Were* there new worlds to conquer in evolutionary biology? Or had E. O. Wilson, a tenured Harvard professor barely into his thirties, seen his best years? On sabbatical in Tobago in 1961, Wilson pondered his fate and plotted his future. After scanning the academic terrain, he set his sights on population ecology.

Population ecology is the study of how the environment—construed broadly to include not just climate and flora but prey and predators, symbionts and competitors—shapes the evolution of a species. Much theoretical progress was yet to be made in the field, Wilson could see, but there was one obstacle: he had no mathematical training beyond algebra and elementary statistics, and quantitative analysis of an increasingly arcane sort was common in ecology. In Tobago he vowed to "lift myself to mathematical semiliteracy"—not just in preparation for the next few years but because he saw, more generally, "the necessity of being analytic in order to encompass and advance the study of particular messy areas." He spent the final months of the sabbatical studying calculus and probability theory with the help of learn-at-home texts. Upon returning to Harvard he enrolled in an undergraduate calculus course, and for the next two years, while keeping up his teaching, research, and writing, he continued to take courses in math.

Even in possession of these credentials, Wilson sought collaboration with a more gifted mathematician, and even before earning them he had chosen his man: Robert H. MacArthur, by consensus the greatest ecologist of his generation. MacArthur, who died of cancer in 1972 at age forty-two, was, Wilson says, "the only true genius I ever met," a man who by sheer force of insight could "convert a swampland into a garden." MacArthur and Wilson decided to look into patterns of population on islands, and Wilson suggested they do it fast. "I beat the drums," he recalls. "I said, 'You know, this is a field that could really open up in this new way.' And he agreed completely. And very shortly we were off and running." *The Theory of Island Biogeography* was published in 1967 and proved seminal.

Wilson and one of his graduate students later conducted an elaborate test of the theory. They picked out an island in the Florida Keys, fumigated it until no insect was left alive, and then observed its reinvasion. The ensuing growth and stabilization of population in the various species roughly accorded with the theory, thus capping what for Wilson was a bracing experience. When he and MacArthur had entered the field of island biogeography, he recalls, "it was just a total mess. . . . In fact, it was described and dealt with in much the same way that a lot of sociology and anthropology is dealt with today—descriptive, verbal terms." Their success in cleaning the field up with rigorous analysis helped give him the courage to confront "a big, sloppy system, like human society."

Wilson loves to name things, and he has a knack for it. He comes up with terms that nicely capture the essence of their referents, or have an air of academic authority, or both. He summed up the tendency of similar species that share turf to evolve in opposite directions as "character displacement." He distinguished between "releaser" pheromones, which immediately trigger specific behaviors, and "primer" pheromones, which work obliquely, altering patterns of future behavior. Among his other coinages: "phylogenetic inertia," "evolutionary pacemaker," "behavioral scaling," "epigenetic rule," "ethnographic curve." Wilson is a lover of islands, and upon realizing that his condition had no name, he created one: "nesiophilia." This inventiveness has served him well. In science, as in commerce, labels are important. Among the gifts that he brought to his several collaborative efforts of the sixties was a flair for packaging.

Toward the end of that decade he began capitalizing on his linguistic proficiency in another way. For the first time in his career, he authored a book solely. And, like the next two books he would write, it was not a technical monograph but a book that would succeed or fail in large part on its literary merit—on the logic of its organization and the grace of its exposition. It was about insects.

Insects had for a long time gone understudied, in Wilson's view, and one reason, he believed, was that the technical literature was in disarray; articles were scattered about in obscure journals, and some were written in obscure languages. The last general, English-language review of the social insects was William Morton Wheeler's book by that name, published in 1928. Further, much of the more recent work on insect societies was not in tempo with new ideas in evolutionary biology; articles were anecdotal and incoherent, devoid of organizing principles. Wilson decided to straighten things out. The result was a large and lucid work—more than five hundred dense, double-columned pages of diagrams, tables, pictures, references, and crystalline prose: *The Insect Societies*.

As remarkable as the book itself—it won universal praise and is still considered by some colleagues his best book—is the fact that he wrote it in eighteen months. The formula for such productivity is simple, he says: you work seven days a week, and you work long days. You sleep short nights. If you're a family man, like Wilson, you work at home whenever possible. "I feel it really is true that work is a central source of meaning for human beings," he says when pressed to explain such behavior. "It's been said many times before in many more eloquent ways that work gives daily satisfaction. And a lot of work, for me, gives me a lot of daily satisfaction."

Wilson barely paused for breath before embarking on *Sociobiology*. The idea of bringing insect, other invertebrate, and vertebrate societies into a single analysis had first struck him in the mid-fifties, after he was deemed the faculty member most nearly qualified to supervise the work of Stuart Altmann, a graduate student who specialized in primate behavior. Accompanying Altmann to observe rhesus monkeys in Puerto Rico, Wilson found himself comparing the social structure of primates with that of insects. And he heard from Altmann a new word: *sociobiology*. It had a nice sound to it. Fifteen years later, as he was finishing *The Insect Societies*, he began to look closely at the literature on vertebrate behavior, and was surprised to find it glaringly deficient; even the innovators, such as Lorenz, had been slow to pick up on useful new ideas, such as kin selection, from outside their fields. Wilson's instinct for order stirred. Perhaps the study of *all social behavior* could be tidied up with a handful of potent ideas.

He spelled out this ambition in the last chapter of *The Insect Societies*, "The Prospects for a Unified Sociobiology." He wrote: "As my own studies have advanced, I have been increasingly impressed with the functional similarities between insect and vertebrate societies and less so with the structural differences that seem, at first glance, to constitute such an immense gulf between them. Consider for a moment termites and macaques. Both are formed into cooperative groups that occupy territories. The group members communicate hunger, alarm, hostility, caste status or rank, and reproductive status among themselves by means of something on the order of 10 to 100 nonsyntactical signals. Individuals are intensely aware of the distinction between groupmates and nonmembers. Kinship plays an important role in group structure and has probably served as a chief generative force of sociality in the first place. In both kinds of society there is a well-marked division of labor . . ."

What began as provocative speculation ended as conviction. "I wanted to have a coda, a concluding chapter with punch in it," Wilson says. "And when I wrote the thing I persuaded myself." He laughs. "That's the way those things happen. I came to feel that there was really quite a future to this. I had a sense of exhilaration. I thought, Wow, we really can put something together that's unifying here."

Some people may have trouble believing that a single moment of exhilaration could have sparked the intensive three-year campaign that would result in *Sociobiology*. But these people have never seen E. O. Wilson get excited about intellectual unity. I have. Once I was talking to him in his office when the phone buzzed. It was his aide-de-camp, Kathleen Horton, informing him that Marvin Minsky was on the line. Minsky had been a fellow junior fellow in the fifties, shortly before he became one of the founding fathers of artificial intelligence. Wilson picked up the phone. "Hello, Marvin. How you doing?" Pause. "Oh, working hard, as always." Long pause. "That's exciting," he said, jabbing his left index finger forward in sync with the expression, as if Minsky were in the room to appreciate the gesture. "Hey, listen, why don't you two come over for lunch sometime and let me show you some ant colonies?" It turned out that Minsky and a colleague were interested in studying the way ants process information. He needed say no more; Wilson took over from there. "A thing that's been continuously running through my mind," he told Minsky, "is that it's now time for people like yourself in AI and those who are looking for new creative approaches in computer design to have a look at the superorganism— you know, not just the brain but the ant colony." They agreed to have lunch, and Wilson hung up with a look of nearly ecstatic delight on his face. "How 'bout that?" he said. "After thirty years, Minsky and I are talking about getting together and working again." He walked over to his refrigerator to retrieve lunch, still marveling. "Sometimes a person like that can bring up one idea or new technique that you've never heard of"—he opened the refrigerator door,

turned toward me, and swept his hand outward to signify vastness—
"and it'll just change everything."

It is hard to say exactly what makes Wilson's enthusiasm so literally
boyish. Partly it is his golly-gee, Gomer Pyle walk, but mainly it is
that the enthusiasm is authentic. E. O. Wilson has somehow preserved the ability to get extremely and unabashedly excited about
ideas, an ability that—in those few people who have it in the first
place—typically begins to fade not long after college graduation. He
likes making connections, and he likes the people he makes them
with. If, in conversation, you discover an intellectual affinity—both
he and I, for example, have trouble understanding things written by
Michel Foucault—his voice warms with fraternity. Sometimes you
get the strange feeling that this fifty-seven-year-old man is the new
kid in school and wants to be your best friend. It wouldn't shock me
if one day his eyes lit up in mid-conversation and he exclaimed,
"Hey, let's build a fort in the woods!" And if he said it, he would
do it.

Wilson did not imagine, at first, bringing humans into his grand
synthesis. Indeed, the last sentence in *The Insect Societies* seemed to
exempt them from sociobiological analysis: sociobiology, he observed,
could be "expected to increase our understanding of the unique qualities of social behavior in animals as opposed to those of man." As a
postscript, he tossed in an ode to free will from Pierre Huber, a
nineteenth-century entomologist: "This great attribute, which signi-
fies unbounded wisdom, induces us to admire those laws by which
providence rules the insect societies and reserves to herself their
exclusive direction; and it shows us that in delivering man to his own
guidance, she has subjected him to a great and heavy responsibility."

What convinced Wilson to subject humans to the explanatory
power of sociobiology was the resolution, at last, of the tension
between his religion and his science. "I had a lot of inner struggle,"
he says. "I don't mean to say that I was tempted to return to a
fundamentalist view or even an essentially Christian, religious view
of the world." But neither could he accept the view, prevalent among
intellectuals, that the religious experience is nothing more than "an
excited mental state," and is thus not in need of special explanation.

He had seen firsthand—and felt—the depth of its appeal; he was certain that religion had biological roots, that at some point it had been good for the genes. The question was how.

During the early research for *Sociobiology*, an answer occurred to him, the answer he would later articulate in *On Human Nature*. Religion, he speculated, congeals the identity of the adolescent and instills a sense of purpose that pays off genetically, fueling his ambition and channeling it toward investment in his future and that of his family and society. The adaptive value of the religious impulse, through selection at the individual, kin, and even group levels, had earned it a place in our genetic heritage.

If something as seemingly inexplicable as religion could be plausibly explained biologically, what other baffling human behaviors might similarly yield to Darwinism? Wilson concluded that the fun was just beginning.

This belief was bolstered by his association with Robert Trivers, one of the most creative theorists in the history of evolutionary biology. During the early 1970s, Trivers showed, among other things, that kin selection is not the only Darwinian way to explain altruism. Something he called "reciprocal altruism" may make genetic sense even among organisms not related to one another—indeed, even among members of different species. Reciprocal altruism sounds like a very commonsensical idea: you do a favor for me and later I do a favor for you. And it *is* commonsensical—so long as the animals in question can, like humans, do favors with the expectation of being paid back. The less obvious thing that Trivers found—and demonstrated with the neat logic of game theory—is that reciprocal altruism could thrive even among organisms incapable of "understanding" the concept of reciprocation. The only prerequisite is that they be able to recognize individual organisms and store information about their past behavior. (Darwin himself, actually, first outlined these conditions for this sort of altruism, but, like many of the other insights he tossed off, this one had to be rediscovered before being developed and fully appreciated.) The meagerness of this prerequisite means that the roots of reciprocal altruism—and of the cerebral mechanisms governing it—could stretch deep into our mammalian past. Trivers, no more cautious in his speculation than Wilson, wrote that "friendship, dislike, moralistic aggression, gratitude, sympathy, trust, suspicion can be explained as important adaptations to regulate the altruistic system."

In addition to showing that familial altruism is not the only kind, Trivers gave theoretical precision to the definition of its limits. He argued that, because members of a family do not have *all* their genes in common, their genetic interests differ, and a certain amount of conflict is therefore "natural." Thus, the same big brother who defends his sister on the playground one day may—with equal fidelity to the interests of his genes—complain the next that too much money is spent on her clothes and not enough on his. Such a complaint is an example of the "parent-offspring conflict" that, according to Trivers, we should expect to see in many species, and that should change predictably over a family's life cycle. This theory held a special attraction for Trivers, who had not had placid relations with his parents, and Wilson was also impressed by its power.

During the early 1970s, Trivers was an assistant professor at Harvard. He read and critiqued every chapter of *Sociobiology* in draft, and his enthusiasm reinforced Wilson's. Wilson needs extra doses of enthusiasm like King Kong needs steroids, but that is what he got. And it showed. The production of *Sociobiology* was intense even by his own standards. "I got excited and just kept at it," he recalls. He worked twelve to fourteen hours a day, seven days a week, for more than two and a half years. "I don't mean I was spending ninety hours a week just on that book," he says in clarification. "I was also carrying out my regular duties at Harvard."

Sociobiology: The New Synthesis was published in 1975. The initial critical reaction was favorable, even, in places, rhapsodic. John Pfeiffer, writing in the *New York Times Book Review*, called the book's birth "an evolutionary event in itself, announcing for all who can hear that we are on the verge of breakthroughs in the effort to understand our place in the scheme of things." Wilson says, "I know some [of sociobiology's] critics—a couple of extreme and nasty critics who couldn't think of anything else worse to say—have said that the book was very successful because Harvard University Press went all out in marketing it. That's not true. They didn't really get rolling until it got some tremendous reviews and started selling like crazy. *Then* they came out with big full-page ads in the *New York Times* and the *New York Review of Books*."

If marketing didn't play a role in the book's initial success, packaging probably did. *Sociobiology* had the look and feel of authority. It had pages ten inches square, each capable of accommodating more than 1,000 words, and it had 697 of them, counting the 22-page glossary, the 33-page index, and the 65-page bibliography. Equations and graphs abounded, as did eye-catching, 2-page illustrations depicting some of the more marvelous, and some of the more human, behaviors in species ranging from insects to primates.

Lest anyone doubt its cosmic significance, the book opened with the obligatory cryptic epigraph:

> Arjuna to Lord Krishna: Although these are my enemies, whose wits are overthrown by greed, see not the guilt of destroying a family, see not the treason to friends, yet how, O Troubler of the Folk, shall we with clear sight not see the sin of destroying a family?
> Lord Krishna to Arjuna: He who thinks this Self to be a slayer, and he who thinks this Self to be slain, are both without discernment; the Soul slays not, neither is it slain.

The first sentence of the first chapter ("The Morality of the Gene") established the author as a man not inclined to shy away from the big issues: "Camus said that the only serious philosophical question is suicide. That is wrong even in the strict sense intended." Within three pages Wilson had mapped out the past and future growth of sociobiology—*literally* mapped it out; two diagrams, one for 1950 and one for 1975, depicted the amount of conceptual turf occupied by such interrelated inquiries as sociobiology, behavioral ecology, ethology, neurophysiology, and comparative psychology. As you might guess, sociobiology had grown; ethology, its nearest rival, had shrunk. But this shift was nothing compared with the change in fortunes that the next twenty-five years would bring, as illustrated by a map for the year 2000. The accompanying text explained, "The conventional wisdom . . . speaks of ethology, which is the naturalistic study of whole patterns of animal behavior, and its companion enterprise, comparative psychology, as the central, unifying fields of behavioral biology. They are not; both are destined to be cannibalized by neurophysiology and sensory physiology from one end and sociobiology and behavioral ecology from the other."

The book had three main parts. The first—"Social Evolution"—
was a five-chapter introduction to basic concepts and theories. The
second—"Social Mechanisms"—was organized by the categories of
behavior that, Wilson argued, possessed coherence and unity even
when stretched across the full breadth of the animal kingdom. There
were chapters on communication, sex, aggression, territoriality, parental care, dominance systems. The final part—"The Social Species"—consisted of ten chapters, each about the social behavior of a
major animal group, such as cold-blooded vertebrates, birds, or nonhuman primates.

The irony, in view of the subsequent controversy, is twofold. First,
there was relatively little discussion in *Sociobiology* of human beings.
Only the last of twenty-seven chapters is devoted to them, and they
appear sparsely elsewhere. Second, what Wilson said about humans
was not generally unreasonable. And, contrary to popular impression,
its upshot was not that people are animals. In placing the human
species in an evolutionary context, Wilson took pains to highlight its
uniqueness. This is done with particular power in chapter 18, "The
Four Pinnacles of Social Evolution."

The first pinnacle of social evolution is occupied by the colonial
invertebrates, such as the corals. Wilson calls these societies almost
"perfect"; the degree of cohesion among individuals, and of their
subordination to the society's welfare, is very high. In some cases,
in fact, such as the siphonophores that together constitute the Portuguese man-of-war, the word *society* seems inadequate. "So extreme
is the specialization of the members, and so thorough their assembly
into physical wholes, that the colony can equally well be called an
organism." As you might suspect, the individuals in these utterly
altruistic societies are genetically identical; the logic behind cooperation, fairly compelling among ants, which share three fourths of
their siblings' genes, here has irresistible power.

The ants, along with the bees and other highly social insects,
occupy the second pinnacle of social evolution. These societies, Wilson wrote, are "much less than perfect." To be sure, altruism is
common, and the societies are closely knit; "the castes are physically
modified to perform particular functions and are bound to one another
by tight, intricate forms of communication." Nonetheless, the insects
are distinct beings; they have an identity independent of the colony,

even if they cannot live long outside of it. Moreover, conflict occurs. Female wasps contend for egg-laying rights, and bumblebee queens attack daughters that attempt to lay eggs of their own. When a queen dies, a struggle over succession ensues.

It is the third pinnacle of social evolution on which egotism is carried to its extreme. Into this category fall all vertebrates except people. Division of labor is seldom apparent in these societies, and "selfishness rules the relationships between members. Sterile castes are unknown, and acts of altruism are infrequent and ordinarily directed only toward offspring." Moreover, no one appears to be having a very good time. "By human standards, life in a fish school or a baboon troop is tense and brutal. The sick and injured are ordinarily left where they fall, without so much as a pause in the routine business of feeding, resting, and mating. The death of a dominant male is usually followed by nothing more than a shift in the dominance hierarchy, perhaps accompanied, as in the case of langurs and lions, by the murder of the leader's youngest offspring."

Note the trend. As organisms become more complex, from corals through ants to baboons, they become less social and more selfish. Note also the beauty of the underlying logic. When individuals are genetically identical, they display almost unlimited cooperation and altruism; when related by a degree of three fourths, they display considerable cooperation and altruism but some independence and selfishness. When the degree of relatedness is merely one half, they display much independence and little cooperation or altruism. Broadly speaking, all of life accords with the theory of kin selection.

There is no reason, *on genetic grounds*, to expect humans to defy the correlation between rising complexity and declining sociality. After all, they, like most other complex animals, are related to their nearest relatives by one half and to most of their neighbors by considerably less than that. Nonetheless, humans *do* defy this correlation. They uniquely mix personal autonomy with intricate social cooperation. "Human beings remain essentially vertebrate in their social structure," Wilson wrote. "But they have carried it to a level of complexity so high as to constitute a distinct, fourth pinnacle of social evolution. They have broken the old vertebrate restraints, not by reducing selfishness but rather by acquiring the intelligence to consult the past and to plan the future. Human beings establish long-remembered contracts and profitably engage in acts of reciprocal altruism that can be spaced over long periods of time, indeed over generations. . . . Their transactions are made still more efficient by a unique syntactical language. Human societies approach the insect societies in cooperativeness and far exceed them in powers of communication. They have reversed the downward trend in social evolution that prevailed over one billion years of the previous history of life. When placed in this perspective, it perhaps seems less surprising that the human form of social organization has arisen only once, whereas the other three peaks of evolution have been scaled repeatedly by independently evolving lines of animals."

This view of humanity may not be especially inspiring, but it is not very dispiriting, as scientific views go. It demonstrates due appreciation of the fact that great intelligence and complex language permit us alone to defy the logic of crude genetic calculus.

Wilson's critics did not dwell on this part of the book. They focused on the last chapter: "Man: From Sociobiology to Sociology." Even here, Wilson's claims were not all that intemperate. He mainly suggested, as had others, that, since humans are products of natural selection, the perspective of the evolutionary biologist could shed light on the nature of things such as aggression, ethics, aesthetics, romance, and religion—that, in other words, the genes play a prominent role in these areas.

Of course, this idea goes against the grain of what for decades reigned as the conventional wisdom in the social sciences. The second college edition of *Webster's New World Dictionary*, published in the early 1970s, defines *human nature* as "the common qualities of all human beings; esp., *Sociology*, the pattern of responses inculcated by the tradition of the social group." The implication is amazing but basically accurate: mainstream sociologists long ignored the possibility that human behavior is significantly influenced by the genes. Thus even the less adventurous speculation in Wilson's final chapter was destined to incur wrath. Still, he could have kept the wrath at a bearable volume by couching his speculation more judiciously. It wasn't so much what Wilson said that got him into trouble as how he said it.

He began the last chapter by trampling on territorial sensibilities: "Let us now consider man in the free spirit of natural history, as though we were zoologists from another planet completing a catalog of social species on Earth. In this macroscopic view the humanities and social sciences shrink to specialized branches of biology; history, biography, and fiction are the research protocols of human ethology; and anthropology and sociology together constitute the sociobiology of a single primate species." Wilson then proceeded to describe human behavior in terms so starkly clinical as to suggest—to anyone hunting for such a sentiment, at least—indifference to the casualties of capitalism. "The members of human societies sometimes cooperate closely in insectan fashion, but more frequently they compete for the limited resources allocated to their role-sector. The best and most entrepreneurial of role-actors usually gain a disproportionate share of the rewards, while the least successful are displaced to other, less desirable positions." Among the chapter's other highlights was Wilson's observation that "deception and hypocrisy . . . are very human devices for conducting the complex business of everyday life." Wilson concedes, "I stated the argument in strong terms to attract attention to it." But, he adds, "I'm not a controversialist."

The controversy began painfully close to home. A Boston-based group called Science for the People, which is located at the left end of the American political spectrum, opened fire in a letter to the *New York Review of Books* after it published the late Conrad Waddington's generally favorable review of the book. The letter linked sociobiology to past variants of genetic determinism and to their consequences, including the Nazi holocaust. It placed Wilson in "the long parade of biological determinists whose work has served to buttress the institutions of their society by exonerating them from responsibility for social problems." Under no circumstances would such charges have delighted Wilson, but they would have been easier to take had not two of the signatories been members of his own department: Richard Lewontin, a geneticist whose office is one floor below Wilson's, and Stephen Jay Gould, a paleontologist. Both, Wilson notes, have said that Marxism has influenced their views on evolution. These influences, as Gould and Lewontin see them, are at a deep level; Lewontin has coauthored a book about natural selection as seen from the prospect of Hegelian dialecticism, the philosophical foundation of Marxism. But some colleagues see more straightforward connections between their science and left-wing ideology: by minimizing the role of genes in human behavior, Gould and Lewontin are free to attribute most of the inequality of material achievement among Americans to an unjust economic system. (It is ironic that many ardent capitalists also profess to believe in the equal native potential of people—and thus feel free to blame poverty on indolence.)

Until he read the letter signed by Gould and Lewontin, Wilson considered them his friends. Lewontin had been brought to Harvard from the University of Chicago, in the face of faculty qualms about the obtrusiveness of his political activism, on the strength of Wilson's recommendation. Gould, Wilson says, had spoken to him days before the letter appeared without mentioning it. Wilson felt, says one friend, that he had been stabbed in the back.

By 1975, American college campuses had begun their long, rightward drift, but they had not yet drifted far enough for Lewontin and Gould to have any trouble finding support. The student newspaper, the *Harvard Crimson*, ran antisociobiology articles, and protestors passed out leaflets during Wilson's lectures. Walking through Harvard Yard, he could hear demands for his dismissal broadcast by bullhorn. "It got very rough in the fall of seventy-five," he says. "These groups were calling me a fascist, a typical example of Harvard's promotion of imperialist, capitalist, fascist ideas. It got so noisy—you know, there wasn't anything directly threatening, but I began to worry that it might start getting physical." Wilson went before the faculty council and asked for a show of support. He got sympathy but no action. "I had a feeling at that time of being really excluded. It was not a comfortable feeling. There were several prominent professors here at Harvard, and substantial numbers of students and outsiders and so on, who were declaring me a very dangerous person and not fit to be on the faculty. You can get a feeling of isolation and exclusion."

Not long into the controversy, Wilson decided to take his case to the people—to write a smaller, less technical book for a lay audience. "Harvard University Press was dying to capitalize on the publicity," recalls Robert Trivers. "And Ed was dying to strike while the iron was hot." Many of the themes in *On Human Nature* were drawn from *Sociobiology*, but in elaborating on them, Wilson made them more palatable, employing a sense for political nuance not previously evident. In a brisk demonstration of rhetorical power, he managed to reconcile his belief in the depth of genetic influences on human behavior with a basically egalitarian, liberal, and optimistic outlook. He argued that universal human rights is a principle with a natural foundation, based on the "mammalian plan," and that homosexuality may also be natural, a product of kin selection. Yet aggression, he wrote, is not as natural as you might think; while it is true that "we are strongly predisposed to slide into deep, irrational hostility under certain definable conditions," there is "no evidence that a widespread unitary aggressive instinct exists."

All told, there was plenty of cause for cheer in *On Human Nature*, and plenty of evidence that E. O. Wilson was not a cold rationalist after all, but a sensitive humanist. "To the best of my ability, I made allusions to literature," he told an interviewer from *Omni* magazine in 1978, the year the book was published. "I've used quotes from Yeats and Joyce and so forth."

CHAPTER

FOURTEEN

CULTURE

The political undercurrents of sociobiology, real or imagined, were not the only grounds for opposition to it. Close inspection of Wilson's writing became a small academic industry—*Sociobiology Examined, The Sociobiology Debate, Sociobiology: Sense or Nonsense?*—and social scientists and philosophers unearthed scores of what they said were serious conceptual flaws. Two of these, in particular, caught Wilson's attention. First was the problem of the missing mind. Conrad Waddington asked in the *New York Review of Books*, "Is it not surprising that in a book of 700 large pages about social behavior there is no explicit mention whatever of mentality?" The other element that didn't seem to fit neatly into Wilson's analysis was culture. If the genes really exercise great control over human behavior, as he seemed to be claiming, then why are different human cultures—whose underlying gene pools differ only slightly, after all—so different? And how is it that in the course of recorded history, a period too short for substantial genetic evolution, cultures have changed so completely? It says much about E. O. Wilson—about his perseverance, his self-certainty, and his almost poignantly earnest reaction to criticism—that in 1981 he came out with a book called *Genes, Mind, and Culture*.

To understand the "gene-culture theory" presented therein, you must understand what the word *culture* means to a biologist: culture is any information transmitted from one organism to another nongenetically. Take all the information you have received from other people, subtract the part that came from your parents in the form of DNA, and what's left is culture. Thus, the fact that you are endowed with teeth is due to genetically transmitted information, but the fact that they haven't yet fallen out of your head, victims of decay, is due to nongenetically transmitted information; brushing teeth is a culturally acquired habit. Similarly, while the capacity—and perhaps even the inclination—to hum is in the genes, the tunes hummed are part of culture.

This definition of culture, though justly lauded for its simplicity and power, has what some people consider an unpleasant side effect: according to it, humans are not the only cultural animals. For example, the white crowned sparrow, if raised in isolation, can muster only the sorriest excuse for a mating call; to become an effective crooner, it must listen to its elders. This rule holds in many species of bird, and in some, such as the saddleback of New Zealand, stylistic quirks have diverged and merged until distinct regional dialects formed. Other birds have other kinds of culture. Oystercatchers, long-billed birds that roam the American shores of the Pacific and Atlantic, get at the interior of a mussel shell in one of two ways: either by hammering it until it breaks, or by sneaking their beaks in and snipping the muscle that keeps it shut. While all oystercatchers are genetically capable of employing both methods, none does. You're either a snipper or a banger, and it all depends on how your parents brought you up.

On a few occasions, animals have been caught in the act of cultural innovation. In the 1950s, at the Japanese Monkey Center, Imo, the resident macaque genius, discovered an efficient way to process mixtures of sand and wheat: after throwing the stuff into the sea, she skimmed the wheat off the top. The technique spread from macaque to macaque, and within five years was prevalent. (This invention came some time after Imo had popularized the preconsumption rinsing of sweet potatoes.)

However humbling it may be to think of culture as something we share with animals no more cerebral than a monkey or a saddleback, the biologist's conception of culture is worth entertaining. Far from devaluing culture, it ends up showing what a watershed the birth of cultural evolution was—and what a meta-watershed the birth of human culture was; it affords a sweeping view of our place in the history of life. And careful consideration of the issue of cultural evolution helps explain why some people think E. O. Wilson would be better off today if he had never listened to the critics and had simply left mind and culture alone.

There was a time on this planet when the only way for a parent to transmit instructions to an offspring was via the genetic code. There still exist vestiges of that time, the most obvious being plants. Once a wild oat has sown copies of itself, it must adopt a hands-off policy.

Animals, of course, are different from plants. Among the differences is that they aren't rooted in one spot all their lives. It turns out that this property—"motility" is what biologists call the ability to get around—is in a way the reason for culture's existence. Given the fact that our early ancestors could move from one place to another, it was all but inevitable—or at least it appears so in retrospect—that the day would come when parents supplemented their genetic instructions with cultural instructions, such as "Wash behind your ears or I'll do it for you!" Basic laws of information processing, in conjunction with natural selection, impelled motile organisms toward culture almost inexorably.

The logic behind the impulsion has been spelled out by John Bonner, a biologist at Princeton University, in the book *The Evolution of Culture in Animals*. Purposeful motion, as we have seen, implies the processing of meaningful information; any organism, even a bacterium, that moves with anything other than aimlessness must have a system of communication that connects it to the environment and— especially in the case of complex organisms—coordinates its various parts during propulsion. So essential is this requirement that motile, multicelled organisms developed chains of cells devoted entirely to the shuttling of messages—"nervous systems," one such organism finally dubbed them. Generally speaking, these informational networks, even if rudimentary, operate more efficiently when centralized. So as soon as there were nervous systems, there was evolutionary pressure toward centralization. Exceptions live on, of course; the neural net of a jellyfish is equally tenuous everywhere. But by and large, evolution has enforced this adage: if it is worth having a nervous system, it is worth putting most of its power in one place, a headquarters from which branch offices can be controlled. That headquarters is called a brain, and when an animal has a brain, it is on the road to culture, because brains are good at processing and storing information.

In their ability to process information—to turn inputs into outputs, reports into instructions—brains are much like the DNA that created them. In fact, what the DNA has essentially done is build a guardian information processor that is faster, more flexible, and more capacious than itself. Both DNA and the human brain exist because it takes information processing to defy the spirit of the second law, but the brain is capable of more facile defiance than is DNA; in the time it takes bacterial DNA to receive information about an energy shortage, order the construction of a flagellum, and be transported to more hospitable climes, the human brain can make a whole series of decisions that are, like the DNA's "decision," aimed at keeping its vehicle intact: put hands in pockets to shield them from the freezing wind, head toward a warm building, and, if possible, make it a building with food in it. (Meanwhile, the brain is unconsciously orchestrating all kinds of intricate internal patterns that keep the vehicle running smoothly, somewhat as bacterial DNA oversees everyday functioning.) These two great organic information processors—human DNA and the human brain—are utterly dependent on one another; without genes, the brain could not exist, and without the brain, the genes would not last long enough to get their information sent to the next generation. It is a case, as Bonner has noted, of true symbiosis.

For a time, the information that brains stored was not culturally inherited. This may sound impossible. After all, didn't we define culture to encompass *all* nongenetic information? No. Culture includes all nongenetic information transmitted *from one organism to another.* There is a third class of information—neither cultural nor genetic—that originates in the environment. It is absorbed (or created, depending on how you stand on some epistemological issues) via firsthand learning. Thus, after eating a poison berry and growing violently ill, an animal might have second thoughts about returning to that particular berry bush. Somewhere in its brain, the animal has stored information that does not reflect favorably on the bush. It has learned something about its environment.

It was a different kind of learning that ushered in culture: learning by imitation. If a young animal pays close attention to the behavior of a parent who has already gone to the school of hard knocks— watching, for example, what is eaten and what eschewed—it can save itself a lot of trouble. Instead of starting its education from scratch, it taps the memory of a graduate. The offspring's offspring learns more efficiently still; it gets the benefits not only of the behaviors that its progenitor learned from its grandprogenitor but also of the refinements and extensions with which its progenitor's experience has tempered them.

Thus, a body of practical information is shipped from one generation to the next, and each generation does a little editing, deleting some obsolete data here and throwing in some original research there. This is cultural evolution—the selective transmission of nongenetic information from organism to organism.

As a means of adapting to environmental change, cultural evolution has one large advantage over genetic evolution—the same advantage that brains have over genes: speed. Consider a bunch of elephants being ravaged periodically by hunters. Any genes predisposing these elephants to run away from loud bangs would probably do well on the evolutionary marketplace, but the chances of such a timely genetic mutation are slim. It is not so unlikely, though, that elephants could have already acquired, through genetic evolution, the inclination to *learn* an aversion to *any* circumstances surrounding the sudden and widespread death of other elephants and to transmit that aversion culturally to their offspring. Thus, elephants that witnessed the carnage might thereafter beat a hasty retreat upon hearing any loud bang, whether from a rifle or a backfiring Jeep, and neophyte elephants might pick up the habit. This effect has in fact been observed; young elephants that had never seen hunters behaved as if they had. In a single generation, cultural evolution had done what genetic evolution would have taken millennia to do.

Imitation has its limits as a means of getting information from one organism to another, and at some point animals began encoding their lessons in symbols. The use of language—not just auditory, but visual, chemical, or tactile—has been observed in dolphins, whales, birds, and bees, and experiment has shown that the lower primates have an impressive capacity for symbolization. So it is possible that in a number of nonhuman species, symbols carry cultural information. Still, only humans have crossed the linguistic threshold, developing a set of arbitrary symbols that, through varied combination, can carry a vast and everchanging array of instructions between the generations. Human parents can say things like "Sit up straight!" and "Don't roll your bread into a little ball!" rather than try to impart these lessons through charades that would look so ludicrous as to end up undermining their authority.

Language, in addition to preserving parental dignity, permits the transmission of bodies of information so complex or abstract that imitation is a hopelessly inadequate means of getting them across anyway: ideologies, theories, novels, religions, laws. These and other fairly abstract elements of culture are continually being tested against alternatives, and the losers are being cast into the dustbin of information, alongside unsuccessful genes. All across the cultural landscape, survival is going to the fittest.

This broad parallel between cultural and genetic evolution—that both consist in the selective transmission of information—has impressed many thinkers over the years, including E. O. Wilson and Charles Lumsden, who made central use of it in their gene-culture theory. But there are differences between the two evolutions that bode ill for anyone who would like to think about the two with equal clarity. To begin with, the units of cultural evolution are more difficult to define than the units of genetic evolution (although deciding where one gene ends and the next begins is not as simple as you might think). Is a song a single, indivisible unit of cultural evolution, comparable to a gene? If so, then what do we say about hybrids, such as Jimi Hendrix's version of "The Star-Spangled Banner," into which he stitched a stretch of "Taps"? And what about unintentional mutations (at age six, I sang "My country 'tis of thee, sweet land of liver tea")? Perhaps it is better to think of the song as comparable to the entire genome—an organism's full complement of genes—and each stanza, or each phrase, as the analogue of a gene. But if so, then what is an entire record album—a population? What are all the Beatles' albums, taken together? What is rock and roll? The more you think about it, the murkier it gets.

And music, actually, is one of the neatest cases of cultural evolution. Everyone agrees on the words to "The Star-Spangled Banner," but do any two people have exactly the same conceptions of Marxism or capitalism? Are any two Southern Baptists equally ascetic? Do any two hedonists indulge in the same things to the same extent?

These questions raise an issue that further muddies the waters. In biology there is a distinction between genotype and phenotype. The gene for blue eyes is the genotype, and the trait resulting from the gene's interaction with the environment—that is, the blueness of the eyes—is the phenotype, or the phenotypic "expression" of the genotype. What, if anything, in culture corresponds to that distinction? Are terms such as *Marxism, dictatorship of the proletariat, Christianity,* and *redemption* mere phenotypes—surface expressions of underlying bodies of ideas that exist immaterially, like Platonic ideals? And if not, then how can we justify considering the currency of these words essentially the same phenomenon as the currency of their counterparts in other languages—counterparts that, after all, may bear no physical resemblance to them? One escape from this difficulty is to claim that in the brain of every Marxist, regardless of nationality, is some sort of generic neuronal pattern representing the Marxist body of information. Or, since this claim has the unmistakable ring of falsehood, one could argue instead that the German and American words for Marxism are bound by the fact that proponents of Marxism *behave* similarly in the two countries—look askance at bankers, await the impending collapse of the bourgeoisie, and so forth. Is the behavior, then, the cultural phenotype? Is the word the cultural genotype? You tell me.

Yet another discommoding difference between cultural and genetic evolution is that the pathways of cultural transmission are perplexingly diverse. Culture can spread, like genes, "vertically," from parent to offspring, but it can also spread "horizontally," from sibling to sibling, friend to friend, singer to audience, writer to reader.

All of this suggests that theorizing about cultural evolution is much harder than theorizing about genetic evolution. And it is. Nonetheless, the parallels between the two have proven irresistible. For decades, population biologists have been simulating genetic evolution with mathematical models that depict genes spreading through a population over thousands of generations. The idea that cultural evolution, its subtlety and snakiness notwithstanding, might be similarly amenable to mathematical treatment has enticed not only E. O. Wilson but also a number of other adventurous biologists into the dicey business of thinking about human culture.

Science itself offers clues as to how cultural evolution could succumb to quantitative analysis. Theories compete with theories much as organisms compete with organisms and genes with genes. And while a theory's influence is more subtle than that of a gene for eye color, some of its manifestations can be tracked with precision. We can, for example, count the number of times the article in which it was unveiled is cited by other scientists. (The *Science Citation Index* saves us the trouble of actually camping out in a library and sifting through journals.) Thus, one measure of how big a splash William Hamilton's theory of kin selection made is the number of citations of his paper, "The Genetical Evolution of Social Behavior," in the years after its appearance in 1964. As it turns out, this theory made no immediate splash. For six years virtually no citations of the paper appeared in the *Science Citation Index* (a fact that doesn't speak highly of the average biologist's awareness and judgment). In 1970, seven articles cited Hamilton's paper, and three years later the number was only eight. Then annual increases became steady and substantial—to seventeen in 1974, twenty-six in 1975, forty-three in 1976, and fifty-four in 1977.

Did the publication of *Sociobiology* in 1975 have anything to do with this late blooming? Did Wilson's body of cultural information enter into a symbiotic relationship with Hamilton's body of cultural information? There is some evidence to that effect. Although citation of Hamilton's paper had begun to accelerate the year before Wilson's book was born, there were no citations in *social* science journals until 1976, the year after. (There were five that year, and the number grew thereafter, exceeding twenty in 1983.) Also suggestive is the fact that Wilson accidentally induced a mutation in the title of Hamilton's paper, and the mutant version proved prolific. In the bibliography of *Sociobiology*, Wilson mistakenly referred to the paper as "The Genetical *Theory* of Social Behavior." (He probably was misled by the memory of Ronald Fisher's 1930 book on population genetics, *The Genetical Theory of Natural Selection*.) Although at least one paper published before *Sociobiology* contained the same error, it was only after 1975 that the error flourished; one survey of the literature, done in 1980, found that more than a fourth of all post-*Sociobiology* citations of Hamilton's paper contained Wilson's mistake.

Another index of intellectual influence—a bit more ambiguous than citation rate and harder to quantify—is the use of terminology. The number of copies of the word *sociobiology* that find their way into books and magazines says something (though it is not always clear what) about the success of the larger work that the word represents. So to gauge in a very rough way the status of Wilson's crusade, you could go to the card catalog, or the *Reader's Guide to Periodical Literature*, and count the number of listings under the word *sociobiology*. Similarly indirect indices could help gauge the influence of the gene-culture theory. It is represented in the vocabulary of science by *culturgen*, Wilson's word for "unit of cultural evolution." Other terms have been proposed for the same role—*meme, idene, sociogene*—and *culturgen* is now in competition with them. *Culturgen*, in other words, is a good example of a culturgen.

After working more than a year's worth of seven-day weeks, Wilson and Charles Lumsden got *Genes, Mind, and Culture* out on the market a few weeks ahead of a book called *Cultural Transmission and Evolution: A Quantitative Approach*, written by two Stanford biologists, Marcus Feldman and Luigi Cavalli-Sforza. Feldman and Cavalli-Sforza developed equations to describe the promulgation of units of culture, such as farming techniques, and tried thereby to sketch some basic patterns of cultural evolution. Unlike Wilson and Lumsden, they made no attempt to depict the relationship between genetic and cultural evolution. In reference to that fact, Wilson has called their approach "timid." Feldman has called Wilson and Lumsden's book "pseudoscience." Cultural evolution, like genetic evolution, is competitive.

The gene-culture theory may not be the most ambitious theory in the history of science, but it's certainly in the running. It addresses, as Wilson said shortly before publication, "one of the grand remaining problems of science. I say that not because I work on it. I work on it—I head in that direction myself—because I believe that. It's difficult to imagine any more interesting or promising a domain for scientific inquiry. The only one I can think of comparable is at the ultimate level of subatomic physics." He paused and added, "Or the ultimate problems of cosmology—you know, where it all came from, how the laws of the universe began, and so on."

In trying to understand the gene-culture theory, it is best to stick with simple examples of cultural innovation, such as the discovery by chimpanzees, at some point in natural history, that tools could improve their daily termite yield markedly. (They stick a twig into a termite nest, swish it around, then pull it out and harvest it as if finishing off a Popsicle.) In cases like this, the two evolutions can affect each other in a fairly straightforward way. If the stick-licking trick makes chimpanzees more likely to survive and reproduce, genetic evolution will favor any genes that incline chimps to learn it; genes predisposing chimps to pick up sticks, or to jam them into rotting tree trunks, or just to imitate other chimps that are good at finding food, may do better than alternative genes. In this way, culture can alter the course of genetic evolution; it can redefine the criteria by which genes are selectively preserved.

But that is not the end of the story. These changes in the gene pool in turn induce cultural change; the more genes predisposing chimps to lick ants off sticks, the more ants will get licked off sticks. Thus, the two evolutions exert reciprocal influence. The result is "coevolution."

As simple as these observations may seem, they form the cornerstone of the imposing mathematical edifice that Wilson and Lumsden construct. For all the obscure symbols that cloud the pages of their book, the basic contours of their model are not hard to discern.

The coevolutionary model's two main ingredients are culturgens and "epigenetic rules" of cognitive development. A culturgen is any fairly distinct body of, or manifestation of, cultural information: a theory, a song, a hairstyle, a habit (the use of soap, say, or of a stone ax). The epigenetic rules are instructions for the brain's assembly inscribed in the genes; they can be thought of as roughly tantamount to genes—or, at least, to discrete *groups* of genes. Some of these rules, Wilson and Lumsden say, are more conducive to the acquisition of certain culturgens than are other rules. And some culturgens, they add, are more conducive to the organism's survival and reproduction than are other culturgens. The rest, it seems, more or less follows.

Consider an example that Wilson and Lumsden do not discuss but that fits easily into their model. It is somewhat like the chimpanzee stick-licking example. Suppose there are two competing culturgens in some early hominid society. One culturgen is to take sticks and stones and knock edible animals over the head with them. The alternative culturgen is to hunt the old-fashioned way, by hand. Suppose also that in the hominid gene pool are two epigenetic rules with relevance to these culturgens. Rule A gives hominids a brain that harbors a fondness for sticks and stones; if you take someone whose brain was built with the help of rule A and confront him with a choice between the two hunting methods, he will have a 90 percent chance of deciding to use weapons. Rule B implies no such fondness and leaves its carriers with a mere 20 percent chance of choosing weapons.

It is safe to say, of course, that before long most hunters with epigenetic rule A will be high-tech hunters, whereas most hunters with rule B will be low-tech. So if high-tech hunters are more likely than low-tech hunters to survive and reproduce, as they probably will be, then hominids possessing rule A will have more children than those possessing rule B, and eventually rule A—or, more precisely, the genes on which it is written—will pervade the gene pool. This fact, in turn, will ensure the predominance of the high-tech hunting culturgen. Pretty soon just about everyone will be picking up sticks and stones and knocking animals over the head.

This is a no-frills example. It doesn't reflect all the assumptions that Wilson and Lumsden make for the sake of realism or mathematical convenience. But it catches the drift of their argument: the two evolutions are intimately connected, every step of the way. As Wilson puts it, genes hold culture "on a leash": cultural evolution cannot proceed faster than genetic evolution will permit; until epigenetic rule A dominates the gene pool, the high-tech-hunting culturgen cannot dominate the culture.

This is not to say that culture is impotent. Cultural evolution can, like a big dog being walked by a little old lady, pull on the leash and drag its master along; the high-tech hunting culturgen, by dint of its contribution to genetic fitness, pulls epigenetic rule A into predominance. But whichever of the two evolutions makes the first move, it cannot go far without the other. "The genes and culture are inseverably linked," Wilson and Lumsden wrote in *Promethean Fire*, a popular version of *Genes, Mind, and Culture* published in 1983. "Changes in one inevitably force changes in the other."

Wilson appears to be treading a fine line here, bordering on genetic determinism. He is not saying that genes determine culture in the strictest sense; rule A leaves *some* possibility that the hominid won't use weapons, and rule B leaves some possibility that he will. Still, the genes do influence the acquisition of culturgens in a very specific way; even if there are not groups of genes that *dictate* the use of weapons, there are groups of genes that strongly encourage it. And only with the prevalence of such genes can weapons proliferate. This point bears repeating: within the gene-culture theory, *genetic change is a prerequisite for significant cultural change.*

After translating this conception of the coevolutionary process into mathematical language and throwing in a few additional assumptions (for example, an individual's choice of culturgens may be influenced by the choices of peers), Wilson and Lumsden are ready to roll. They can introduce a new epigenetic rule and a corresponding culturgen into a hypothetical population and then, after the resulting ripples in the gene pool and the cultural fabric have died down, gauge the effect on both layers of information.

From these mathematical experiments they derive principles that they deem fundamental. The "amplification law" says that even slight genetic mutations can, given time, bring dramatic cultural change—and, therefore, that two populations whose gene pools differ only slightly could have very different cultures. The "thousand-year rule" says that, when pulled along by cultural evolution, genetic evolution can move a significant distance in only fifty generations—and, therefore, that the rapid cultural change over the past millennium could conceivably have had some genetic underpinning.

With the formulation of these two principles, Wilson and Lumsden disposed of one of the large, nagging criticisms of sociobiology that had driven Wilson to coevolutionary theorizing in the first place. Even if genes affect human behavior specifically and pervasively, according to their model, human cultures *could* differ greatly both from place to place and from time to time. The problem of culture had been felled with a few deftly wielded equations. It was almost like magic.

The thousand-year rule, if right, means that since the beginning of rapid cultural evolution, some 30,000 years ago, there has been plenty of time for genes to "track culture"—for important cultural innovations to foster parallel changes in the gene pool. Thus, genes may to this day bias people toward certain culturgens and away from others. "This is not to say that every nuance of culturally transmitted behavior in prehistory was hardened genetically in the form of very specific epigenetic rules," Wilson and Lumsden cautioned in *Genes, Mind, and Culture*. Still, "the results of our model do suggest that time has been more than adequate for substantial coevolution and the establishment of some degree of epigenetic bias in virtually every category of cultural behavior."

It is easy to think up examples that seem to support this contention. Like fires. My father taught me how to build a fire. He transmitted the instructions culturally, and without them I would be a poorer fire builder—or, perhaps, no fire builder at all. Yet there may be something more than culture at work here. When I build a fire, and gaze into it as it grows, I fall into something like a Neanderthal trance; my mind glazes over, and I feel a visceral serenity. Given the undoubtedly important role of fire in keeping our ancestors warm and safe on cold and scary nights, it is conceivable that "Neanderthal trance" is indeed an apt term. The trance, in other words, may be genetically built bait, designed by evolution to entice me into building fires night after night and discourage me from wandering very far from them. And maybe without such bait I would have had no interest in acquiring the fire-building culturgen. I may be a walking example of the kind of link between genes and culture that Wilson has in mind.

Unfortunately for the gene-culture theory, few things in modern human culture are as elemental as fire. Today, most culturgenic choices (Wilson selected the word *culturgen* partly because of its "reasonably graceful adjectival form") are of a kind that did not confront our distant ancestors: whether to listen to jazz or rock and roll; whether to take the plane or the train; whether to put it on Visa or American Express. If the acquisition of all culturgens had to be specifically facilitated by genes, how could the granddaughter of someone who never saw a television decide not only whether to buy a VCR but which brand to buy? For that matter, how could science progress? How long would it take us to get from culturgens like "gravity" to culturgens like "relativity" if we had to wait for the gene pool to catch up? How long would it take a word like *culturgen* to win acceptance if acceptance required the proliferation of mutant genes facilitating its comprehension and pronunciation?

These questions point to a fact that any theory purporting to capture the connection between the two evolutions must somehow accommodate. In modern human societies, cultural evolution seems to be moving along at a much crisper pace than ever before, but genetic evolution seems not to be. The number of scientific and technological advances per year, the speed with which fashions change in music and clothes, and with which jargon comes into and falls out of vogue—all of these appear to have grown over the past few centuries. Yet there is no evidence that change within the gene pool has accelerated commensurately, and it may, in fact, have slowed down; one of its traditional spurs—widespread death before the age of reproduction—is now absent.

A poet might find irony in the prospect that cultural evolution is speeding up just as genetic evolution is slowing down: What perverse fate has kept them from flying along ever faster together? Any such poet would have entirely missed the point: ever-more-rapid cultural evolution is the *cause* of the deceleration of genetic evolution. The things that in modern societies have put an end to widespread death before the age of reproduction include penicillin, vitamin pills, cheap heat, new and improved fertilizers, eyeglasses, appendectomies—all of which are fruits of science and technology and therefore of cultural evolution. So our poet would be well advised to give up on that particular irony angle and ask instead questions like this: Has genetic evolution given birth to a second great evolutionary process and then died at the hands of its offspring? Has cultural evolution assumed a life of its own?

These are tempting, if slightly mystical, metaphors and, for Wilson's theory, dangerous ones. For if genetic evolution has left us with minds broad enough to absorb an ever-accelerating flow of culturgens, so that culture can race ahead while the genes stand still, then things look bad for any theory that makes cultural change dependent on genetic change.

It is not surprising, then, that Wilson resists the idea of such "broad minds." He and Lumsden belittle the notion that people have merely a "capacity for culture"—that we are the passive agent and culture the active one.

But the fact is that broad minds needn't be passive or indiscriminate. Consider the affinity for explanatory economy that attracted

Wilson to the theory of natural selection and Fredkin to his theory of digital physics. Suppose this affinity had been "hard-wired" into the brain by evolution during early human attempts to understand the environment and manipulate it usefully. Once wired in, why would it have to change? After all, it has served well in choosing ideas ever since Aristotle—and not just scientific ideas; explanatory economy guides the auto repairman, the police detective, and seekers of misplaced keys the world around. And a slightly different but basically congruous penchant for economy shapes our sense of beauty in design; we like floor plans that accommodate many rooms with a few clean lines, board games that generate much complexity with a thin rule book, and the Ronco Vegematic, which slices, dices, chops, grinds, grates.

It is not hard to imagine other broad, hard-wired criteria for judging culturgens. But we can save time by saying that many of them fall under the heading "rational faculty"; people can tell, by and large, what works and what doesn't work. They hang on to the things that work the best and discard the rest.

Work best at doing what? What is it that people want culturgens to do for them? The answer, surprisingly, is not that complicated. There appear to be basic, universal, human goals in the light of which culturgens everywhere are judged. Almost all people, for example, want to eat, to have a secure place of residence, to amass resources, to be highly thought of, to meet members of the opposite sex, to kiss them on the lips, etc. They rationally select the culturgens that get them closer to these goals.

In a way, the view of cultural evolution contained in the previous three paragraphs is very sociobiological. After all, both the ability to evaluate culturgens rationally and the goals in whose light they are evaluated are products of evolution, brought to us by epigenetic rules that were naturally selected. But, unlike Wilson's version of epigenetic rules, these rules wouldn't have to change much over the centuries; rationally pursuing these goals is still, after all these years, a good way to get genes passed on. So, according to this revisionist sociobiological view, rapid cultural evolution doesn't presuppose rapid genetic evolution. The culturgens change, but the rules of thumb remain the same. Wilson could have said things like this in *Genes, Mind, and Culture* and remained faithful to *Sociobiology*. He didn't *have* to talk about a "leash principle" to stay within a Darwinian framework. But he did.

The present speed of cultural evolution may sound like a shaky basis for challenging the gene-culture theory. After all, Wilson would concede that his model of coevolution doesn't mirror the dynamics of *contemporary* cultural evolution—wouldn't he? Yes, now that the gene-culture theory has been widely and sharply criticized, he is willing to make that concession. And he now says, further, that the "leash principle" shouldn't be taken *too* literally. But he and Lumsden didn't make either of these qualifications in the book. And though most of the book's examples of culture come from premodern societies, such as the Yanomamö of Venezuela, at one point he and Lumsden unveil elaborate equations meant to describe changes in women's fashion in the eighteenth, nineteenth, and twentieth centuries.

They don't, it turns out, argue that the gene pool had to change for these fashions to evolve, so the point of fiddling around with the fashion example never becomes entirely clear. But the example does have some unintended value: it illustrates how hard it is to model contemporary cultural evolution and provides some clues as to why. A graph accompanying the example depicts three quantifiable elements of fashion—the levels of the neckline, waist, and hemline—as they changed between 1788 and 1936. For most of that time, these numbers behaved stably, inching up for decades and then down for decades. Indeed, the book's analysis is premised on their following cycles of about a hundred years. What Wilson and Lumsden do not note is that shortly after 1900 these numbers go into gyrations, rising or dropping more steeply than ever before and reversing themselves just as dramatically.

What had happened to upset the even flow of culture? There's no way of knowing, but it is likely that the technology of cultural transmission had simply advanced. It was not until the final years of the nineteenth century that photographs could be reproduced by printing press. (The requisite technology—the halftone engraving—is familiar to newspaper readers even today: degrees of shading lie in densities of dots.) This capacity to illustrate advertisements and articles with real pictures is considered by some historians responsible for the rise, around the turn of the century, of the mass-circulation magazine, which offered cheap subscription and relied for income mainly on advertising, thus ushering in the great advertising boom. In 1900, $542 million was spent on advertising in the United States. In 1920 the figure was $2.9 billion. And that was just the beginning. In 1920, regularly scheduled broadcasting began on radio. Television came two decades later.

Today, thanks to these and allied technologies, fashions no longer creep slowly from New York to the Midwest before becoming passé; they leap across the continent, even as west coast innovations fly eastward in cross-fertilization. So too with jargon, music, radical chic, neoconservative waves, and the rest. All move like lightning along the national nervous system.

What this means is that the term *cultural acceleration* fails to capture the magnitude of what is happening. The very structure of cultural evolution is changing, and the change is as stark as the difference between vertical and horizontal; the intergenerational continuity of culture is being broken by intragenerational bonds of information. Today a fifteen-year-old, upper-middle-class boy in suburban Texas may have more in common with a similarly situated boy in California than either has with his father. The two teenagers listen to the same music, wear the same kinds of clothes, watch the same TV shows, use the same slang, and have the same problems—for starters, an inability to relate to their parents. It is probably not overstating the case to say that the generation gap is in large part a product of information technology. And this gap restricts the power not just of parents but also of E. O. Wilson's gene-culture theory. For it speaks of a cultural evolution that is moving much faster than even an elastic leash would allow.

Richard Dawkins, the Oxford zoologist who was writing *The Selfish Gene* even as Wilson was putting the finishing touches on *Sociobiology*, devoted the last chapter to cultural evolution. Though Dawkins's view of life is in many ways closely aligned with Wilson's (*The Selfish Gene*, really, is an extended essay about the most powerful theories in sociobiology), his ideas about culture are quite different from Wilson's—and, indeed, from those of many people. Dawkins has taken the commonsensical view of cultural evolution and turned it inside out.

He considers units of cultural information—"memes," he calls them, not "culturgens"—to be replicators; they are best seen, he says, not as passive units of information, being shipped from one organism to the next, but as living things that, like the genes, actively exploit their environment. To a meme, a brain is only a means to an end; it is a temporary nesting site, good ground for reproduction. "Mary Had a Little Lamb" is one of the most prolific parasites in history.

Dawkins thinks of brains as the primordial soup in which memes first spawned, and he thinks of computers, newspapers, and other such media for storing or transmitting nongenetic information as modern soups. Thus, you might say (though Dawkins doesn't) that the idea of putting salt on food found fertile ground in the brain's genetically based weakness for salty tastes and easily jumped from parent to child, parent to child, parent to child—until it ran into a formidably hostile body of information about high blood pressure, a body that employed the latest in memetic reproductive technology (TV, radio, printing presses) to launch a mass assault on the salt meme and vanquish it from many craniums.

A common response to this sort of inverted depiction of the relationship between genes and memes is: "Well, sure, you can look at it like that without stretching the truth too much, but what's the point?" The point is that you gain insights that might otherwise elude you forever. Central among these is that what is good for the memes may not be good for the genes. To be sure, many, perhaps most, memes *do* contribute to the survival and reproduction of the genes. Washing hands before dinner has probably prevented a certain amount of disease, and medical science has forestalled millions of otherwise early deaths. Even clothes, in addition to keeping people warm in the winter, serve to advertise their socioeconomic class, their tastes, and how seriously they take themselves; to the extent that such things matter in marriage, clothes may save time in the mating game.

But some memes perform no such services. Some sneak into our brains by making us feel good when they are not really good for our genes. Overindulgence in salt is an arguable example, heroin addiction a less arguable one. Religion is a more problematic example than either. On the one hand, the idea that there is some purpose to life, or that another life awaits us after the earthly one is over, may prevent an existential crisis that would induce prereproductive suicide, and may keep postreproductive parents happily providing for their children and thus for their genes. On the other hand, the moral strictures of the Church may keep men from just the sort of philandering that could expand their genetic legacy. Certainly there are a number of priests and nuns whose genes are not eternally grateful to the Christian body of information.

The Shakers, a religious sect that flourished in America during the eighteenth and nineteenth centuries, were the battleground for an especially brutal clash between memes and genes. The Shakers forbade procreation, marriage, and, for that matter, physical contact between the sexes. So, unlike conventional Christian memes, Shaker memes could not exploit genetic paths of transmission and travel conveniently from parent to offspring. They had to be powerful enough to attract and infest a continual influx of converts. For a time, they were; at one point more than a dozen Shaker communities were thriving in America. That Shaker memes nonetheless faced an uphill struggle is reflected in the dearth of Shakers today; the sect is essentially extinct. (The extinction was partly economic in origin. The Shakers supported themselves by crafting and selling the simple but elegant furniture for which they are still noted, and the industrial revolution, by cutting the cost of manufactured goods, priced them out of the market.) It would be interesting to see whether today, with the help of modern communications technology, a sect prohibiting procreation might have better luck in horizontally transmitting enough memes to flourish in spite of clogged vertical conduits.

The fact that memes need not help the genes of their host organism, and may, in fact, kill that organism, seems at first to represent a sharp difference between genetic and cultural evolution. But careful thought reveals memes to be in this respect *exactly* like genes: in the end, they are in this thing only for themselves. Somewhat as warning-call genes casually discard a prairie dog while keeping copies of themselves in circulation, heroin-addiction memes will, after using a teenager as a spawning ground and sending copies of themselves to his friends, leave him by the wayside without the slightest regret. In both cases only the fittest information survives, and in both cases fitness may or may not coincide with the welfare of the information's carrier.

One appealing aspect of Dawkins's view of cultural evolution is its resonance with the tempo of modern life. The idea that memes, or culturgens, are in some sense alive, and are reproducing as fast as they can, is consistent with the impression that these days information literally assaults our senses; from radios, televisions, magazine stands, subway placards, "portable" cassette players that dwarf the people carrying them, the memes leap out at us. Sometimes modern life seems to be a random and absurd juxtaposition of symbols, all battling for access to our brains. (And, as we just saw, the fact that they have won past battles does not mean they are good for us.)

There is irony in the fact that these symbols bring change in the circumstances of our lives, and that the increasing ease of their transmission brings accelerated change. As we've seen, the information age, in its earlier years, brought hopes of conceptual order in the social sciences. Claude Shannon and Norbert Wiener showed how information, the medium of social interaction, could be quantified, and they provided basic concepts, such as feedback, for its analysis. But, however promising the intellectual undercurrents of new information technology, its social effect has been to accelerate change so dramatically that, even if valid generalizations about society were forthcoming, they would probably be obsolete not long after their articulation. The information age has made human society more comprehensible in principle and more inscrutable in practice, clearer from afar and murkier up close.

FIFTEEN

MUTINY

Genes, Mind, and Culture contained, by Wilson's reckoning, the first conceptual structure of its kind, an unprecedented combination of rigor and scope. Its chains of impressively alien mathematical symbols stretched all the way from molecular biology to anthropology and sociology, spanning the chasm that for so long had separated the social and the natural sciences. "We've *built* the bridge," Wilson told me one day in the spring of 1981, shortly before Harvard University Press published the book amid much fanfare. The book itself did not inspire widespread agreement with that assessment.

The first inauspicious sign came from *Time* magazine. In 1977, *Time* had given Wilson its seal of approval with a sympathetic cover story on sociobiology. *Genes, Mind, and Culture* was heralded with a less flattering article about Wilson and Lumsden called "Sociobiology's Vaudeville Team."

Most scholarly reaction to the book fell somewhere between lukewarm and rabidly hostile. Richard Lewontin called it "a severe tactical blunder," a revelation of sociobiology's shallowness. Marcus Feldman dismissed some of its central findings, in particular the thousand-year rule, as sleight of hand. "The book starts from the conclusion that you have to explain the way in which genetically controlled behaviors evolve quickly," he said. "That's the conclusion. And now you design your mathematics to prove that." Feldman considered this "a weird way to do science."

Lewontin's criticism had all the unpredictability of Republicans denouncing Democrats, and Feldman's was surprising only in its severity. More notable was the fact that judgments not much less harsh came from people with no conspicuous axes to grind. *Science* magazine, which is to science what *Variety* is to show business, published an article about the book in May of 1981—"Cultural Diversity Tied to Genetic Differences." The author had talked to more than a dozen biologists, and their appraisals of the gene-culture theory were mostly negative. "A number of people well placed to comment on the work declined to go on record with their criticisms," he wrote. "A disturbingly loose fit between the model and the world it is meant to reflect was, however, the principal theme." Among the few who were willing to speak for quotation was John Maynard Smith, a British biologist who, while long harboring reservations about *human* sociobiology, has made basic theoretical contributions to sociobiology as a whole and is commonly identified with the field. "It's not that the theory is racist or sexist or anything like that," he said. "I just believe theirs is a simplistic concept of the cultural process."

The problem, as Wilson diagnosed it, was that people didn't understand. It wasn't so much that they didn't understand the gene-culture theory—though that was part of the problem—as that they didn't understand science. Science, he says, is a continual oscillation between expansion and compression. The cycle begins when a scientist—a "systems builder," more specifically—comes along and claims he can compress a large body of information into a small theoretical package. The gene-culture theory, for example, is an attempt to account for masses of data about human cultural history with a fairly small set of principles. It may not immediately be clear how a theory is going to accommodate all the information in its domain; being new, it is still rudimentary. But as it is tested and refined, it will begin to "unfold into a full display of rich detail."

In the meanwhile, though, you can't expect the theory's creator to sit around contemplating its shortcomings. That is left to the "bookkeepers," the scientists who "fill in the blanks." It is their job to collect lots of data and see whether they really do fall into the pattern predicted by the theory.

If you're a bookkeeper, says Wilson, "you can afford to play by the rules and express lack of confidence in your results and express the severe doubts that the experimental data may engender." Wilson used to do a lot of bookkeeping, but these days he sticks mainly to synthesis and systems building. And when you're a systems builder, he says, "you have to have something like faith. You have to believe as you go on that in fact there is some major organizing principle that remains to be discovered. You have to believe that indeed it exists and that no matter how imperfect a formulation may be from one year to the next, or what setbacks might occur, how many seemingly intractable methodological problems originate, that this will prove to be correct."

This is the kind of faith Wilson has in the gene-culture theory. He doesn't expect other scientists to share it, but he does think the theory deserves more patience than it has received, more time to evolve. "We're talking about where biochemistry was in 1850 vis-à-vis curing cancer or telling what the structure of the gene is."

Even before *Genes, Mind, and Culture* disaffected John Maynard Smith and a few others generally regarded as sociobiologists, there had been murmurs of discontent from within the ranks, signs of dissatisfaction with Wilson's conduct as chief publicist and titular leader. Many of the murmurs had originated at the University of Michigan, in the offices that flank the Museum of Zoology. There, over the past few decades, the connection between genes and behavior has been pondered at length by a number of well-known biologists. One of them is Richard D. Alexander, a square-jawed midwesterner whose twang sounds at home with phrases like "pissin' against the wind," which is what he says E. O. Wilson is doing.

Alexander has known Wilson since the 1950s, back when both men were strictly insect biologists. Like Wilson, he subsequently took up the study of human beings, and, like Wilson, he believes strongly that social scientists will remain mired in confusion so long as they study human beings without reference to the process that created them. But he isn't the adroit writer that Wilson is, and he labors in the relative obscurity of Ann Arbor. His books—*Darwinism and Human Affairs*, for example—don't sell like Wilson's books sell. Alexander never seems to find himself in the limelight, where E. O. Wilson always seems to be.

To begin with, says Alexander, Wilson does not deserve to be known as sociobiology's founding father; in fact, sociobiology does not even deserve to be known as sociobiology. All of the ideas that form its foundation were in place before 1975, when Wilson applied that label to them. They traditionally had fallen under the broad rubric of evolutionary biology, which, Alexander believes, is a perfectly fine rubric for them to fall under. Further, they had come from people such as Trivers and Hamilton, not Wilson. And even if Wilson did arrange these ideas attractively and shower them with publicity, he muddled some of them in the process, according to Alexander. Here, for example, is Wilson's introduction to kin selection:

> Imagine a network of individuals linked by kinship within a population. These blood relatives cooperate or bestow altruistic favors on one another in a way that increases the average genetic fitness of the members of the network as a whole, even when this behavior reduces the individual fitnesses of certain members of the group. The members may live together or be scattered throughout the population. The essential condition is that they jointly behave in a way that benefits the group as a whole, while remaining in relatively close contact with the remainder of the population. This enhancement of kin-network welfare in the midst of a population is called *kin selection*.

Alexander's complaint is that this passage, especially the next-to-last sentence, makes kin selection sound like a form of group selection—as if a prairie dog's death had to benefit an entire group to qualify as kin selection. In fact, though, if two prairie dog brothers, both possessing "altruistic" genes, are about to get devoured by a bunch of coyotes, and one dog suicidally sounds an alarm that saves the other, the gene responsible for that act of altruism has been preserved by it. Thus, Wilson's description of kin selection—and his decision to place it in a chapter called "Group Selection and Altruism"—bore the seeds of confusion. So says Alexander, and he is not alone. Dawkins and several others in the field have taken Wilson to task in print for mishandling kin selection. (Alexander also accuses Wilson of being a bit facile in confidently attributing the social cohesion of ants, bees, and wasps mainly to the unusually high degree of relatedness among siblings; he says that, although kin selection almost certainly played a role in the evolutionary integration of these societies, entomologists in growing numbers doubt that this was the critical factor.)

Alexander thinks Wilson has simplistic ideas about a number of other issues in sociobiology as well; mix such misunderstanding with political naiveté and an instinct for provocation, and you have a public relations disaster. For example, it is a commonplace of biology that *nothing* is "genetically determined." Every trait is determined by the joint influence of the genes and the environment; even genes so relatively straightforward as the ones for blue eyes cannot express themselves without favorable conditions. Alexander claims that Wilson blurred this point in a debate with the anthropologist Marvin Harris on the Dick Cavett show. "Cavett opened the program by saying, 'Look, let's see, this whole thing is about whether human behavior is genetically determined, isn't it, Dr. Wilson?' You know, instead of Wilson saying, 'That's a naive way to put it,' or 'That's not a good dichotomy,' or 'Let's don't be confused starting out that way,' he said, 'Yes, there are some of us who believe there are certain human behaviors that don't vary or that are universal because they have a very firm genetic basis and cannot be changed.' And then he used facial expressions as an example, and Marvin Harris tore into that in a way that I thought was appropriate. I was just sorry that Ed didn't start his whole discussion in an entirely different way. I wanted to be either Wilson or Harris, but if I had been Harris, I think I could have made Wilson look a lot worse than Harris did in that debate."

But Alexander wasn't Wilson and wasn't Harris, and that brings us to another of his complaints—about Wilson's tendency to position himself in the spotlight. In *Genes, Mind, and Culture*, Wilson and Lumsden depicted the gene-culture theory as pathbreaking, if not epoch-making. In an encyclopedic, six-page table, they listed past theories of cultural evolution and assigned to each a capsulized evaluation—the upshot of which was that nothing heretofore had succeeded in doing what the gene-culture theory did. Alexander's work, for instance, "foreshadows but does not express an explicit theory of gene-culture coevolution." Feldman and Cavalli-Sforza had made a "step in the right direction." Also reflecting Wilson's belief in the gene-culture theory's preeminence was his decision to come up with a new term—*culturgen*—for the basic unit of culture. Dawkins's term, *meme*, was already gaining favor in some circles, and its endorsement by eminent scientists on both sides of the Atlantic might well have standardized this bit of terminology once and for all. Why Wilson confused things by throwing in another entry, Alexander can't understand. Nor does he see the need for *epigenetic rule*; it's nothing more than a new name for an old concept, he says. "Ed is very clever and an exceptionally fine giver of names to biological phenomena. Some have stuck and become a very solid part of biological language. Sometimes they don't stick. Sometimes he seems to try to make something appear new when it really is not."

In the end, though, what condemned *Genes, Mind, and Culture* to a cool intramural reception was not its imperialistic overtones so much as its substance. Never before had Wilson's work come so close to genetic determinism. Although in *Sociobiology* he foreshadowed the "amplification law"—"the multiplier effect" was its name back then— he had not gone so far as to suggest a genetic bias in "virtually every category of cultural behavior."

Alexander has a theory about how Wilson wound up sounding like a genetic determinist. From the beginning of the sociobiology controversy, he says, Wilson made the mistake of letting his critics define the terms of the debate; rather than dismiss their distortions as such, he tried to defend the positions wrongly ascribed to him. In *Sociobiology*, Wilson had written, quite reasonably, "It is meaningless to ask whether blue eye color alone is determined by heredity or environment . . . [because] obviously both the genes for blue eye color and the environment contribute to the final product." But, says Alexander, as Lewontin and Gould and others obscured such evidence of Wilson's subtlety and cast his position in cruder terms, Wilson turned into the very straw man they depicted. "In parallel with other biologists who had experienced the same predicament," Alexander has written, "he seems to have *become* a genetic determinist, by defending the phrase and the kind of meaning his 1975 statement about blue eyes denies."

Alexander believes that the word *sociobiology* is now permanently tainted with shades of genetic determinism and that its meaning has been further blurred by association with Wilson's personality; some social scientists, for example, think of an especially virulent strain of academic imperialism when they hear the word. It is, says Alexander, in insisting that the word can maintain its intended meaning—as a discipline, not an idiosyncratic theory—that Wilson is "pissin' against the wind." Alexander, for one, refuses to use Wilson's word—and goes to great lengths in doing so; instead of saying "sociobiologists," he says things like "people who approach human behavior from an evolutionary standpoint."

Some other sociobiologists—Maynard Smith, Dawkins—have also avoided describing themselves as such. Even some people who owe their careers to the word *sociobiology* now shun it. David Barash, a psychiatrist at the University of Washington, cashed in on the socio-biology controversy in 1977 with *Sociobiology and Human Behavior*, his first book. He published a second book in 1985 called *The Caveman and the Bomb*, a look at how our genetic endowment bodes for the prospects of averting nuclear war, and a third in 1986, *The Tortoise and the Hare*, about genetic and cultural evolution, respectively. Bar-ash's outlook remains sociobiological, but the word *sociobiology* does not appear in the index of either of his last two books. Indeed, it is Wilson's critics who lately have done the most to keep the word in circulation. Almost all scholars who set out to minimize the role of genes in behavior—such as Anne Fausto-Sterling, author of *Myths of Gender*—use *sociobiology* to label their caricature of the alternative view. And it is often Wilson's writing out of which they weave the cari-cature.

Meanwhile, the sociobiological perspective continues to pervade the culture. The study of animal behavior in the light of such ideas as kin selection has mushroomed, and the prevalent view of human behavior has moved toward the last chapter of *Sociobiology*. Whether this movement has been propelled more by intellectual merit or by conservative political currents is a subject of debate, but there can be little doubt that the movement is real. When the book was published, the idea that genes might underlie some of the observed behavioral differences between men and women was seldom advanced in polite society—and almost never in print. Since then several national mag-azines, *Newsweek* among them, have run long stories on the biological basis of such differences, and now the idea of innately differing dispositions is creeping into the conventional wisdom. Meanwhile, *Psychology Today* and other popular science magazines routinely run articles about the evolutionary logic behind the strategies we employ (often subconsciously) in romance, at work, and at play. Even Phil Donahue, almost a caricature of the liberal-minded, good-hearted, sensitive guy, named his book about human nature *The Human Animal* and leaned further toward the nature side of the nature-nurture debate than was once considered proper in his circles. Sometimes it seems that everything about sociobiology is flourishing but the name.

Cardinal Mazarin, on his deathbed, is projected onto the space above the blackboard. Wilson is pointing toward the image with open palm. "The great prelate, advisor to Louis the Fourteenth, is dying in 1661, ostensibly without *pro*duce, without producing any replicates of Mazarin DNA." He pauses. "But is he? In this painting we find him surrounded by his nieces and their husbands, who have been raised to high places in the French court and who are prospering mightily and having nephews and grand-nephews and grand-nieces for the cardinal."

It is, of course, an illustration of kin selection, of the fact that genes can wind their way circuitously into the future. Cardinal Mazarin will have no direct descendants, yet his progeny—even if they don't bear his name—will be many. He is in that regard like a young prairie dog who dooms himself by warning siblings of impending danger, or a bee that dies on a kamikaze mission that saves the hive, or an ant whose abdomen explodes in the heat of battle. All of these seemingly sacrificial acts have a subtle yet substantial legacy.

Wilson introduced the slide by saying, "Let me use this famous painting by Paul Delaroche of Cardinal Mazarin on his deathbed to illustrate the current stance of what might be called conventional sociobiology." I had to go back and listen to this sentence on tape a couple of times. On first listening, I read too much into it: I thought he meant that Cardinal Mazarin was supposed to represent sociobiology itself.

CHAPTER

SIXTEEN

GOD

Class is over, and Wilson is standing up front, nearly surrounded, taking questions one at a time. A psychology major wants to know more about the intersection of sociobiology and cognitive psychology. "It's a *rap*idly developing field," Wilson tells her. Where can she find further reading? Well, ironically, there aren't many good books on the subject. Only one comes to mind: *Promethean Fire*, by him and Charles Lumsden. She thanks the professor humbly and walks off. Wilson is ready for the next customer, a man in his thirties sporting long, curly hair, cowboy boots, jeans, and a black ski jacket with the Head trademark stitched onto it. He offers to exchange books with Wilson. Apparently this man has written on Buddhism or Taoism or something, and he's curious about the relationship between evolutionary theory and Eastern philosophy. Wilson expects action on this front very soon: "I think the time may be right. You know, we're reaching a critical mass . . ."

The crowd is slow to thin out, but Wilson is patient, and everyone gets all the time they need. The last remaining student seeks permission to take the final exam late. Wilson steers him over to the first row of seats, where they can talk in private. "Let's see if we can't work out some kind of a compromise," he says. They do, apparently; the student is abundantly grateful.

Now only Wilson and I are left in the lecture hall, and we begin the walk over to his office. On the way, we stop at a lunch truck, where I pick up a ham sandwich. Wilson isn't having anything to eat. He says he's watching his weight. (This is strange; he doesn't have any weight to watch.) On second thought, he'll have a giant cookie to go with his cup of coffee; he sometimes rewards himself with something sweet after lectures.

The bulletin board outside Wilson's office looks like the display window of a bookstore. It features the covers of *Biophilia*, *Promethean Fire*, *Genes*, *Mind*, *and Culture*, *On Human Nature*, *Caste and Ecology in the Social Insects* (which he coauthored with George Oster, a mathematical biologist), and *Sociobiology* (both the abridged, paperback and unabridged, hardcover editions). There is also a poster—the kind you might find in a Hallmark card store—of a frog perched precariously out on a limb. The poster says, "All progress has resulted from those who took unpopular positions."

Wilson is now curator of the insect collection that originally attracted him to Harvard, and his office is adjacent to the Museum of Comparative Zoology. The office is a museum in its own right, and he will with obvious delight take you on a tour if you display the faintest interest. A wooden table near the middle of the room is devoted to the most human of social insects: *Atta*. The ants—which crawl through a network of eleven clear plastic boxes connected by soft plastic tubes that say "Standard for Processed Milk Service"— are busily growing fungus on brownish-yellow stuff that looks like foam rubber from a very old pillow. The freezer compartment of the refrigerator in the corner contains, in vials and jars, essences of ant pheromones. The adjacent counter is covered with large vials holding captive ants and with plastic tubs with labels like "Brazil: Fuzenda Dimona VIII-85" and "TEXAS: Houston Pheidole Moerens Wheeler Col #1." On the wall above the specimens are five black-and-white portraits of great entomologists, including William Morton Wheeler. Over by Wilson's desk, which once belonged to Wheeler, two more pictures hang: a caricature of Charles Darwin that appeared in the September 30, 1871, issue of *Vanity Fair*, and a portrait of Herbert Spencer, who envisioned a day when the sciences would be unified under the all-embracing principle of evolution.

Sociobiology's "archives," as Wilson calls them, are in filing cabinets just beyond the desk. All around is sociobiology's "library"— hundreds of books, many of which contain Wilson's word. About twenty of these are his own, in translation: *Da Natureza Humana*, *Sulla Natura Umana*, *Sobre La Naturaleza Humana*. *Sociobiology* comes in four languages, and the Japanese version comes in five volumes. In an adjoining room are more archives, more of the library, and long shelves filled mainly with journals, including the four, founded in the late 1970s, that have *sociobiology* in the title or subtitle. Actually, that number is now down to three. *The Journal of Social and Biological Structures* changed its subtitle from *Studies in Human Sociobiology* to *Studies in Human Social Biology* in October of 1981, about six months after the publication of *Genes, Mind, and Culture.*

The cinder-block walls in both rooms are off-white. The windows have dark, vertical blinds that, at almost any setting, shut out quite a bit of light. There is an unvarying hum—the heating or ventilation system, or maybe just the fluorescent lights—and it gives the place a feeling of hivelike insulation that is at once cozy and scary.

I am sitting at a small table, not far from the agricultural ants, preparing to eat my ham sandwich. E. O. Wilson is trying to make me feel at home: paper plate, napkin; Would I like some water? Without warning, inspiration seizes him; "I know!" he says, pointing a finger in the air as if he had just discovered the one equation that sums up everything. He walks over to the refrigerator, pulls out a can of V-8 juice, and holds it up for me to see. I accept.

Wilson sits down across the table from me, takes off his boots, and props his white-socked feet up on a chair. The boots, which he just bought at Sears, are the kind construction workers wear, and they're killing him. He's breaking them in for a trip next week to Costa Rica, where he will gather ants. He and a colleague, Bert Hölldobler, are now writing the definitive text on ants, a long-overdue replacement for William Morton Wheeler's *Ants*, published in 1910.

As for mind and culture and coevolutionary theorizing: Wilson has had enough of that for a while and is leaving it to Charles Lumsden, who is now at the University of Toronto. "He's got a whole research group down in Toronto, and it's spreading rapidly as a research effort. It's finally catching on. It took a few years for this to get rolling." I ask how the theory is doing in America. "In the United States the main opposition to it has been—" He stops and thinks. "Let's see, what have the main objections been?" He pauses again. "My impression is that it hasn't been objected to so much as it just hasn't been picked up on yet. People who have really thought it through and looked into it have been fairly favorable." Wilson anticipates growth in this population.

What I want to talk about now is the fate of the word *sociobiology.* Is it indeed being boycotted by some of the top people in the field?

But I hesitate to ask. Throwing this question at E. O. Wilson out of the blue seems a little like walking up to a casual acquaintance and saying, "So, I hear you have a terminal disease." I will try to sneak up on the subject.

I remark (in the spirit of the gene-culture theory) that some people—entertainers, writers, politicians, academics—seem to spend as much energy on cultural proliferation as most animals spend on genetic proliferation. "I believe you're right," he says. "I think there is a kind of druglike surrogate." The distinction is between "gaining immortality through your family and direct children, and maybe even the welfare of your tribe vis-à-vis other tribes" and gaining a "considerably more abstract and tenuous feeling of continuity through your work. I think it's generally true—I hope it doesn't sound too airily idealistic—to think that most scientists have that feeling. You know, even if perhaps their identity gets lost, absorbed into the mainstream—you know, the identity of whatever they did—they get a lot of satisfaction out of feeling that they've permanently changed something."

With what is meant to be a casual air, I say, "Contributions to the vocabulary of science can live on for some time. I think you've coined a lot of terms that have become a permanent part of the language."

"Yeah, that's right," he says. "I take some pride in that." What are some of his most enduring contributions? He thinks for barely a second. "*Character displacement* is there solidly," he says. "And the two types of pheromones—I didn't invent the word *pheromone*, but I used the terms *primer pheromone* and *releaser pheromone*, which are now standard. And *sociobiology* itself—I didn't invent the term, but I selected it from among the alternatives." And "I think I played a role in establishing it as *the* word to use."

This is the moment I've been waiting for, and now, having reached it smoothly, I manage to seize it about as gracelessly as possible: "Are there, uh, are there people doing sociobiology and not calling it that?"

"Yes," Wilson says without hesitation. Clearly this is something he's thought about. And he's decided that this glass, like all of his glasses, is half full. "I think it's a very good sign. It's commonly said that the success of a science is judged by the quickness with which its founder is forgotten. And I noticed the first several years, when I was referred to rather excessively as the father of sociobiology, that

wasn't disputed—when it seemed to be under very heavy fire. Now
that it's really caught—catching—on, I've noticed a certain irritation
on the part of some of my colleagues at my being identified as the
founder, even to the extent that some of these people don't use the
word." Wilson finds most of the terms used in its place either cum-
bersome or inappropriate; *behavioral ecology* (the term of choice at
Oxford) is his favorite, and even it has flaws. Nonetheless, he wel-
comes alternative terminology. It is a "sign that people are wanting
to move on, even to the extent of redefining the discipline, defining
new disciplines, finding new expressions. I think that's a sign of
success."

Given his druthers, though, wouldn't he like to see the *word* live
on as well as the ideas? "Sure. Of course. I have a personal investment
in wanting to see that word last. I thought it through very carefully—
quite apart from the identification I have with it personally—and I
think it's the correct term to use for the discipline as defined." And
don't get him wrong; he hasn't given up on the word. "There's some
resistance on the part of social scientists to pick it up explicitly because
it still has a kind of risqué flavor to it. But that's breaking down, and
I see the word *sociobiology* used more and more now in the social
sciences."

Talk turns to information. The legacy of what Wilson calls "the
Shannon-Wiener era" is mixed, he says. On the one hand, extending
the ideas of Claude Shannon and Norbert Wiener from telephones
and thermostats to kidneys and courtship rituals proved difficult. The
former are simple, while the latter are not. So, after a flurry of
activity, including some embarrassingly naive leaps from the technological to the organic, interest in the interdisciplinary use of information theory and cybernetics cooled. (It is hard these days to find
people who will describe themselves as cyberneticists, and the ones
who will are, more often than not, a bit on the mystical side; the
moral they draw from Norbert Wiener's work is that everything,
ultimately, is connected by information with everything else, and,
like, feeding back off of it, you know?)

On the other hand, Wilson notes, a lot of scientists *are* studying
cybernetics, whether they know it or not. "Knowledge of control and

pattern formation and the reproduction of pattern and so on, at all levels—cells, tissue, society, insect societies—has progressed to the point where, by implication, information is regarded as central," he says. "You'll find the cell biologists and the people in tissue assembly talking about the control of morphogenesis, and that means, you know, information. Similarly, people working on populations are very interested in negative feedback control." So, "although the formal measurement of information, involving bits and transfer rates and so on, has gone out of fashion since the fifties and sixties, we are thinking more strongly about the actual mechanism by which information is accrued and transmitted in living systems. We have better ideas about how it actually happens."

One thing that impresses Wilson about the concept of information is how vividly it demonstrates the power of natural selection. When you actually add up the information in a strand of DNA, or in a human brain, "it changes your perspective entirely. People are astonished at just how much complexity—read 'information'—has been built up in the course of evolution."

This is what Wilson thinks is most important about the revolution in information technology: it permits us to sustain this trend by spectacularly transcending the biological constraints on information storage and transmission. The size of any one person's memory is limited by the genes, and the only way to exceed this limit is to store information outside the cranium. Humans have been doing this for millennia, but not always with great facility. These days, though, we can instantaneously store and retrieve information and rapidly convey it to other people who might be interested in it. "And once you begin that, then I don't think it's just a truism to say that the potential becomes almost unlimited. And that's essentially what the information age consists of—the stepwise improvement in information gathering, storage, retrieval, and transfer." Wilson closes his good eye and rubs it deeply and methodically. It is only mid-afternoon, but on a dark December day, three o'clock feels like five.

Science well illustrates Wilson's point. In 1665, back when a printing press was a huge thing and a huge investment, there was one English-language scientific journal: *Philosophical Transactions*, published by the Royal Society of London. Now that anybody with a few thousand dollars can declare himself a publisher, there are tens of thousands of scientific journals, and they foster a very fine division of intellectual labor. The average scientist doesn't carry any more information than the average scientist of three centuries ago, but today that information is confined to a narrower and deeper field, and there are now many more such fields. "And when you put all that together," Wilson says, "you have a truly impressive superorganism."

A superorganism, needless to say, that is becoming ever more interdisciplinarily integrated, growing more organic not just socially but conceptually. I'm still not sure how literally Wilson believes in reductionism, so I pose the proposition in its most extreme form: Can the linkages among the different levels of organization be made so solidly that someday we will be able to deduce laws of social behavior from laws of chemistry? "Now let me think about that." He stares at the space a few feet over my head for a full ten seconds. "No, I don't think it can be done all the way up the line in the sense of using physical chemistry to describe social behavior." He stops suddenly, gets the look in his eyes that previously preceded my getting a free can of V-8 juice, and, once again, holds up a finger. "Oh, that reminds me of a funny cartoon. I don't know if you've seen it. I've got the original one. It's this guy Harris, you know, who does the cartoons. It's exactly apropos of what you said." He walks across the room and brings back a large piece of white posterboard. The cartoon depicts a professor at a blackboard, on which is written an equation that shows the sum of various chemical symbols, such as H_2O, to be the word *aggression*. He is saying to a colleague: "The answer to sociobiology: sociochemistry." I chuckle woodenly, and Wilson, knowing insincerity when he hears it, hastens to reaffirm the cartoon's relevance. "You know, 'cause that's the question—if social systems can be reduced to biology, why not go ahead and reduce them to chemistry?"

He has an answer: because things are very, very complicated. Phenomena that cannot be anticipated—"positional effects" and other sources of uncertainty—get in the way. "So you cannot, with a knowledge of organic chemistry, or even with a general knowledge of macromolecules, predict exactly the, you know, DNA coding and translation and so on, although once it's worked out, then you say, 'Oh, yeah, that's chemistry, and you can explain that by the laws of chemistry,' and so on. But no chemist would have sat down and predicted it with a lot of detail."

Only too late do I realize that the conversation has gotten out of hand. I've gotten E. O. Wilson on the subject of intellectual integration, and there's no stopping him now. "And in that sense—and this seems so obvious to me that I keep being astonished that people, including a lot of critics of sociobiology, don't see it—we have to go through a constant cycle"—he moves his hands around each other like one of Smoky Robinson's Miracles—"a whole advancing front of cycles of reduction and synthesis often, you know, carried along almost simultaneously, where you're looking at patterns, important patterns of behavior, perhaps social organization, and your ultimate goal is to understand that behavior more deeply than before. So the first step you take—the way the human mind works, and I think it's probably pretty close to the way nature actually is—is to start breaking it down into its component parts, then explaining the qualities of those component parts in relation to each other by more general laws working at the next level down. But your aim is not then to just leave it there, but to reconstitute the pattern you started with in a far more rigorous way. I think that's the way science works. But I doubt that it very often would be the case that when you reduced you went down more than one level of organization."

There is a full second of silence. An opening? It's closed before I can exploit it. "Just to take a concrete example, you can't explain an ant colony, the particularities of an ant colony, caste or positional effects and so on, by going two levels down—say, to the macromolecular structure. That's almost three levels down. But you can explain it by a knowledge of individual behavior and principles of how individual dyads—two individuals, three individuals, four—can relate to one another according to degrees of kinship, common interest, and so on. Thus you can explain a society, these complex patterns, then step back a few meters"—he pulls his head back like a person suddenly gaining perspective—"and see fuzzily an advancing army ant colony as—well, it's an entity in itself; it has particular properties that are worthy of description in their own right. And you can explain that by recourse to understanding organismic biology. But you could not explain it by understanding macromolecular structure. On the other hand, you can understand a great deal of cell structure by recourse to the study of the macromolecular structures."

In short, scientists should be both reductionists and holists. And they should be aware that complete, top-to-bottom reduction, though possible in principle, will never be practical.

We talk some more, and then the phone rings. Wilson walks over and answers it. "Are you having a hard time, dear?" he says with heartfelt, almost paternal sympathy. It's his wife. Snow has begun to fall, and she's having trouble driving, or something like that. Anyway, the upshot is that Wilson has time for only one more question.

Time to roll out the big one: the meaning of life. In both *On Human Nature* and *Biophilia*, Wilson wrote about a philosophical crisis facing humanity. The problem, as he sees it, is that people have a deep-seated, genetically based *need* to get wrapped up in some sort of religious fervor—but science, as luck would have it, is relentlessly undermining religious convictions; evolutionary biology has shown the creation myth to be just that, and Wilson is confident that our moral codes will ultimately be explained in terms of genetic imperatives. Fortunately, he has a straightforward solution: as long as we've got these orphaned religious impulses lying around, we might as well hitch them to a belief system that still has legitimacy—science, for instance. He would like to see people get their epiphanies the way he's gotten his—by becoming engrossed in the "endless unfolding of new mysteries"; by investing their faith not in Genesis but in "the evolutionary epic." We must cultivate a "scientific humanism" that taps the energy of our innate religious drive, says Wilson.

Maybe so, but it strikes me that there is one thing missing from this equation. Religion traditionally has imparted a sense of purpose, a sense that there is some point to living, something we are here to do. It has thus provided reason to struggle against base impulses and existential despair, to try to live with compassion, restraint, and dignity. The evolutionary epic doesn't seem to help much in this regard. The knowledge that we are all related to bacteria makes it no easier to swallow the harsh facts of hard work, brief retirement, and death. How can scientific materialism give meaning to our lives?

"Yeah, you've touched on the subject that's been my main concern," he says. "And that is purpose pursued with energy and enthusiasm." He concedes that the issue is a tough one, and that he doesn't yet have the answers. But he knows they will come through rational, empirical analysis—"in somehow understanding the religious impulse as central to human behavior in a way that not only explains it naturalistically but harnesses it." And he knows the answer will depend on building more bridges, further integrating the body of human knowledge. "I think scientists and theologians have a lot more to say to each other in talking about the sources of the religious drive and the hunger for religious thought. . . . I'm rather hopeful that liberal theologians are going to see this as an interesting new area of thinking and investigation."

It's happened again; he's gotten on the subject of intellectual unity. "There has to be a reconciliation," he's saying. "What's the ultimate form? I'm not so sure. I don't know whether it could end up with a strengthened, more open-ended form of theology and religious scholarship and thought, or whether it will just further diminish the strength of formal religion." He's just warming up. Soon the words are coming out in waves that will later defy attempts at punctuation. "I've always felt that maybe the former would be the case, because if you could marry a formal enterprise of religious thought of scholars, thinkers, you know, exceptionally sympathetic people, the kind that are attracted to religion, marry their activity as specialists on central rites of passage, and the central, you know, empathic operation of human society—you know, themes of the majesty of the species, the future of nature and so on—marry that with an open-ended, more scientificlike inquiry—the source of moral behavior, in order to make moral reasoning more scientific, more sciencelike, without, however, going to the obvious extreme of a simplistic form of materialism— you know, man as machine—that religion could come out of this a winner."

In short, religion can survive as a coherent body of information if it is willing to put up with substantial editing.

Like E. O. Wilson, I was brought up a Southern Baptist. Like him, I encountered the theory of evolution as a teenager. Like him, I was bowled over by its power and beauty. Like his religious faith, mine did not survive this encounter with science in good shape.

But there is one difference between Wilson and me. He seems to have had no trouble filling the void. I, in contrast, regularly get wistful about the days when the question of purpose was settled once and for all, when I knew for certain why I was here and how I was supposed to behave. And somehow I find it hard to believe that he never does. So I ask him: Doesn't he long for the days when he believed there was a God up there watching over him? Doesn't he lose any sleep over life after death? He shakes his head firmly. "None," he says finally and proudly. "I don't worry about my own immortality."

Still, a funny thing happened a couple of years ago. Harvard was honoring Martin Luther King, Sr., and Reverend King, as part of the festivities, was preaching at the Harvard Memorial Chapel. Wilson, being a southerner, was invited to the service. There was a large turnout. The reverend preached fervently, and the congregation sang richly, and one of the hymns hit home with Wilson—"one of the good, old-timey ones that I hadn't heard since I was a kid." Partway through it, E. O. Wilson—scientific materialist, detached empiricist, confirmed Darwinian—started crying.

As if in atonement, he has a perfectly rational explanation. "It was tribal," he says. "It was the feeling that I had been a long way away from the tribe."

PART IV

SOCIAL CEMENT

CHAPTER

SEVENTEEN

WHAT IS COMMUNICATION?

The cellular slime mold is a borderline case. It sits on the fence between society and organism—or, rather, jumps perpetually from one side to the other, leaving doubt as to where it belongs. At birth, the little slime mold cells are so independent that they just barely merit the collective label "colony." Each slides along on its own, engulfing the nutritious bacteria it finds amid the crumbling leaves and rotting wood of the forest floor, more or less indifferent to the fortunes of its peers. But if food becomes scarce, bonds materialize. Millions of cells cluster together and form a tiny slug—about as long as a dime is thick—that then begins inching its way toward any heat or light. Hours later, upon arriving at its destination, the slug halts and reaches upward, as if trying to become the world's first vertical slug. As the effort continues, a bulbous head blossoms; by the time the slug has completed its upward extension, it looks like some sort of miniature, grotesque flower. The cells forming the stalk soon die and harden, but cells at the top, having turned into spores, become the next generation of slime mold cells; hundreds of thousands plop onto the forest floor and begin the life cycle anew, dividing and subdividing with abandon. Cellular autonomy has been restored and will last until the next round of hardship.

There are people—not many, maybe, but some—who like to argue about whether the cellular slime mold is a society or an organism. Regardless of their position on the issue, they never win, and that is probably just as well. The value of the slime mold is that it defies categorization. It comes from one of evolution's twilight zones. It is a remnant of that threshold when cells came together to form organisms. It is a priceless chunk of natural history, fortuitously frozen.

This is not to say that the descendants of the slime mold went on to become snails and squirrels and people. So far as we know, our own cellular ancestors underwent the transition between society and organism in the sea. Nonetheless, it is likely that the evolutionary logic behind the periodic coalescence of the slime mold is the same logic that bound our once-independent ancestral cells into the distant precursors of us. And, for that matter, it is likely that this logic also motivated, in large part, the integration of more recent precursors of humans into the precursors of human societies. At these two different levels of organization, the impetus for integration appears to be essentially the same.

If we can understand this recurring logic behind organic coherence, then we will have a clue as to the logic behind the origin of communication—communication among cells, communication among people, communication in general. For coherence, as Norbert Wiener noted, entails communication. Systems that defy the spirit of the second law tend to talk.

In the case of the slime mold, the communication is chemical. The first few cells to suffer starvation send out vaporous bursts of a substance called acrasin. About fifteen seconds after sensing one of these pulses, a fellow cell will emit a pulse of its own, conveying the message to cells farther down the line, and then move toward the source of the first pulse. Meanwhile, its own transmission is received by this source and reciprocated, so more guidance is on the way.

As it turns out, the symbols that thus cement the slime mold colony are the same symbols that have been defending the integrity of bacteria since time immemorial; in some species of slime mold, such as *Dictyostelium discoideum*, the acrasin is made of cyclic AMP, the molecule that, in the bacterium *E. coli*, represents a dearth of carbon and stimulates a search for friendlier environs. Actually, for all the evolutionary distance traveled between bacterium and slime mold, the cyclic AMP hasn't changed much in meaning. It still says, "We have stumbled into a harsh environment, lacking in sustenance—time to activate emergency plans."

The dependence of the slime mold's coherence on symbols doesn't end after it crosses that fuzzy line between society and organism. After all, the cells in the slug must communicate to preserve the harmony that unity demands. So too with the cells that are us. Back before we crossed the same threshold, back when we were not organisms but societies floating in the sea, our cellular constituents, presumably, sent signals through the salt water. Today they still do; symbols flood the salt water that bathes our cells, coordinating (in conjunction with nervous impulses and other forms of information) the cellular division of labor and embodying the meaningful information that cells need in order to adjust to the body's changing demands. These symbols are hormones, and they do everything from keeping blood sugar constant in the face of fluctuating sugar intake (insulin is the name of this hormone) to mobilizing the energy needed to fight, or flee from, playground bullies (epinephrine—adrenaline— is the name of this one).

Epinephrine, though isolated in 1901, was not really understood until the 1950s, when Earl W. Sutherland, Jr., an American pharmacologist, showed that it induces liver cells to reach into long-term energy storage and retrieve glucose, which can then be burned for a short-term boost. Sutherland also showed that the hormone does not have this effect directly. It transmits its message only as far as the cell's surface, which then relays the information to the interior of the cell via a "second messenger." This second messenger is—you guessed it—cyclic AMP; it is secreted within the liver cells as soon as epinephrine molecules bind to them, and it sets off a chain reaction that ends in glucose. (The DNA is not directly involved in this chain reaction, and thus is not, strictly speaking, the program that converts the input, cyclic AMP, into output, glucose; what the DNA has done, in essence, is to construct a surrogate program—a chemical medium that, on behalf of the DNA, does the conversion.) Cyclic AMP is a common second messenger in humans, and it means different things in different contexts. In this case its meaning, as received from the epinephrine and faithfully transmitted, happens to echo its meaning in bacteria and slime molds: "emergency."

The cellular slime mold is strikingly reminiscent of an ant colony. In both cases a cohesion of mysterious origin turns out to rest on tangible strands of information, every bit as physical as cinder blocks and carburetors. (The same is true of human society, although we, being in the thick of things, don't often pause to appreciate the mystery. We don't realize how inexplicable our collective cohesion would appear to a Martian scientist with a powerful telescope, who could only guess about the patterns of photons and sound waves that orchestrate our daily associations.) The slime mold parallels the ant colony in another way, too. Its communication system, while answering the question of organic coherence in one sense, leaves it unanswered at a deeper level: What is the evolutionary logic that drives discrete organic entities into social, and sometimes even closer, cohesion? Granted that slime mold cells must communicate in order to cohere, why must they cohere in the first place?

In the case of ant colonies, a very plausible answer to this question turned out to rest heavily on the theory of kin selection. Even cursory consideration suggests that in the case of the slime mold this explanation is still more plausible—about 33⅓ percent more plausible. Since the slime mold cells are genetically identical, altruistic cooperation should, all other things being equal, make that much more sense than it makes for worker ants, which are related to their siblings by a degree of only three fourths.

Of course, all other things are not equal, so we shouldn't rush to embrace this explanation of the slime mold's periodic coherence. The evolution of cooperative behavior can be driven by a variety of things ranging from common threats (a hostile climate, say, or ubiquitous predators) to the prospect of greater productivity (such as efficiencies flowing from joint food gathering or joint reproduction). Indeed, these circumstances can be so compelling that cooperation sometimes makes genetic sense for entirely unrelated organisms. Even if the slime mold cells shared no genes, they would have nothing to lose by clustering together during hard times, assuming the alternative was death. So kin selection is by no means the sole explanation for social cohesion in the slime mold, or in any other species.

Still, kinship can only strengthen any existing logic behind cooperation, especially cooperation that has an altruistic element. Thus, the "willingness" of slime mold cells in the stalk to die without reproducing would be harder to explain if their genetic information were not carried by spores to the next generation.

The logic, then, runs as follows. Because of various environmental threats and various potential benefits, cooperative cellular behavior makes some evolutionary sense. Because the cells are so closely related, it makes even more sense. Because this cooperation requires communication, a language evolves. In the case of the slime mold, cyclic AMP, which happened to be handy, was pressed into service as a lifesaving symbol. Hence, the origin of language, in two easy steps.

"Origin of language?" the skeptical reader should now be saying. "You call an aerosol emitted by cells *language*?" Well, yes. In fact, the case for calling acrasin language is strong enough so that skepticism should come from the other side, too—from people who insist that language materialized earlier, in the bacterium, when cyclic AMP first acquired a symbolic dimension. The cellular slime mold, these people will say, has only taken intracellular language and turned it into intercellular language.

And maybe they're right. But even if they're wrong and the first group of skeptics is right—even if neither intracellular nor intercellular communication qualifies as language, even if the only things that qualify as language are messages like "Hot enough fer ya?"—I am willing to bet that language owes a large debt to kin selection, that one of the main reasons our mammalian ancestors started talking was because they spent much time close to very close relatives.

One of the most elemental of mammalian expressions—the warning signal—appears to be a product of kin selection. Even some of our distant phylogenetic relatives, such as the prairie dogs and meerkats, warn one another about predators, and these behaviors almost certainly evolved as a way for genes to look after copies of themselves. In our nearer relatives, the nonhuman primates, warning signals are also common, and in some species it is the warning signal that most closely approximates human language in the fineness of its articulation. The vervet, an arboreal African monkey, uses at least four different "words" to identify enemies; a mammalian predator, for instance, is heralded with a *nyow*, while a large bird elicits a *rraup*.

Warning calls are not the only altruistic signals, and thus are not the only signals that stand a good chance of owing their origin partly to kin selection. Another kind of information that primates tend to share liberally is about sources of food. A small group of chimpanzees, having happened upon a fruit tree, will go into an exuberant series of gymnastics known as the "carnival display"; they run around, whoop and bark, and bang sticks on tree trunks until other members of the troop get the message. If a chimpanzee has just obtained some fresh meat (by, say, kidnapping and killing a baby baboon), other chimpanzees may walk up and beg, extending an open hand and whimpering softly. Sometimes the hunter will oblige. Another favor that one primate will do for another is groom it—stroke its fur, removing unsightly and possibly harmful insects and debris. Some primates, such as the white-handed gibbon, ask for grooming vocally, grunting or squeaking as they sprawl out and offer their fur.

Again, it should be stressed that kin selection is not the root of all altruism. There are other evolutionary reasons to share food and information about food, and to do other favors that entail talking. Still, it seems likely that in small primate troops of twenty to fifty individuals the average degree of genetic relatedness was high enough so that the same genes were frequently on both the giving and receiving ends of altruism. Kin selection very likely greased the wheels of cooperation and thus accelerated the evolution of language.

Once the groundwork for vocal language was laid by kin selection, much of the actual construction was probably underwritten by culture. The finer the nongenetic information a primate could absorb, the more richly it could supplement its genetic behavioral programming, so evolution naturally favored the cerebral equipment for understanding language. Further, the ability to *transmit* such information would also have a high evolutionary value, so long as the recipients were sons, daughters, brothers, or sisters—or, to a lesser extent, nieces, nephews, or cousins. Even after culture became the driving force behind the genetic evolution of the linguistic hardware, much of that force was mediated by kin selection.

Some phases in the evolution of a capacity for communication—including some of the early, premammalian phases—have nothing to do with culture *or* kin selection. Take cooperation for the purpose of genetic recombination—or, as such cooperation is also known, sex. The first sexually reproducing pair of organisms must have in *some* sense communicated to coordinate the fusion of genetic material, and thus has it been ever since, with new systems of signals getting invented along the way. The reason cockroaches are so eagerly slaughtered by D-Con Lure and Kill, for instance, is that it contains sex pheromones. As animals evolved into more elaborate forms, sexual signaling came to involve additional senses, such as sight and hearing. Peacocks engage in self-advertisement that would put Norman Mailer to shame, male stickleback fish swim in a zigzag pattern when in the mood for love, and humans go through all sorts of songs and dances to attract and hold the attention of suitors and suitees. Sex also gives rise to a second kind of signaling, as members of one sex "communicate" about access to members of the other sex; various threat gestures are employed in the animal kingdom as a prelude to (or a substitute for) duels over matters of the heart.

Signals can acquire similarly adversarial overtones when used to define pecking orders. If a brash young chimpanzee is feeling his oats and decides to threaten a higher-ranking chimp, he will unleash a loud bark, though perhaps from a safe distance. If conceding his subservience, he will emit a low-pitched sound known as a pant-grunt and then crouch abjectly before his acknowledged superior. (An expression that looks remarkably like a human smile also serves in chimp societies as a gesture of subservience.)

The extent to which language grew out of kin selection, as opposed to the selection surrounding mating rituals, dominance hierarchies, and other such things, is of more than passing interest, for it bears directly on the question of what communication is. When animals convey messages about food or predators to their kin, they are serving, essentially, as extended eyes and ears, passing on information about the environment uncorrupted. Granted, it may occasionally make evolutionary sense to deceive a brother or sister, since siblings don't share *all* their genes, but this temptation is nothing compared with the incentive for dishonesty in communication between entirely unrelated organisms. Here deception is often the whole point of the message. Animals routinely exaggerate their vigor, strength, and fidelity when seeking a mate, and their ferociousness when competing with other animals for food or turf. Human beings have turned bluster, bluff, and insincerity into sciences all their own.

This somewhat cynical view of communication has found voice in Richard Dawkins and John R. Krebs, coauthors of a paper called "Animal Signals: Information or Manipulation?" Krebs and Dawkins note that what is commonly called communication can also be seen as a very economical way for one organism to affect the behavior of another organism. "A male cricket does not physically roll a female along the ground and into his burrow," they write. "He sits and sings, and the female comes to him under her own power." Communication, seen from this perspective, is simply a way for an animal to "exploit the senses and muscles of the animal it is trying to control."

In a way, this exploitation was inevitable. As soon as organisms have information-processing systems—that is, as soon as they qualify for the term *organism*—they are vulnerable, because such systems confer tremendous power on incoming information. An *E. coli* bacterial cell, for example, has placed considerable reservoirs of energy at the hands of a few cyclic AMP molecules; a slight increase in their concentration triggers a fairly complex sequence of behavior that includes the construction of a new appendage and ends with relocation. So if a unicellular organism wants to influence an *E. coli* cell, it can save itself a lot of time and trouble by controlling the information that has so much power rather than resorting to brute strength. And natural selection finely attuned as it is to questions of economy, is bound to smile on this approach to manipulation. Thus do beetles of the species *Atemeles pubicollis* effortlessly steal food from wood ants; by synthesizing a pheromone that typically comes only from needy fellow ants, a beetle will convince an ant to generously regurgitate its painstakingly collected food. The beetle then eats the food and, without saying another word, departs.

Substituting the word *manipulation* for *communication* should not be taken to imply that there is no such thing as "honest" communication. Warning calls, as we've seen, are often painfully honest. Still, the purpose of a warning call is not to further the cause of truth but to get a relative to behave in a manner that protects its genes; like many other messages, a warning call is just a way for genes to efficiently steer vehicles containing copies of themselves. The fact that this is best done honestly is incidental. When such steering calls for dishonesty—such as claiming that Santa Claus likes only those children who eat their vegetables—dishonesty is usually forthcoming.

So Charles S. Peirce was deeply right when he said that the meaning of an utterance is the behavior it leads to. And *pragmatism* was indeed an apt term to associate with language. It so happens that under some circumstances—communication within a cell or an organism, or among closely related cells or organisms—the pragmatic purpose of language can be realized with accurate information. But under other circumstances it can't. The cyclic AMP molecule that some innovative bacterial cell sent to its DNA a couple of billion years ago was perfectly honest in its representation, but once the doors of perception were opened, it was just a matter of time before other organisms began sending in misleading signals. Information, in the end, is influence; and communication is control.

Another way of making the same point is to say that once organisms begin communicating with each other, the line between reports and instructions begins to blur. Thus, if you call someone on the phone and ask how he is, and he replies that he's very busy, this message is at once a report and an instruction. It may be a faithful representation of reality, but it is also a way of subtly suggesting that you not stay on the phone for too long. Its possible accuracy as a report does not make it any less an instruction, or any less a manipulation of you. Similarly, when a boss lets it slip in front of a sometimes recalcitrant subordinate that he has been inundated with résumés from people willing to work long hours for low pay, he may be telling the truth; but he may not be, and in either event he is issuing an implicit threat, and an instruction: quit griping and get some work done.

So once organisms start communicating, the definition of information gets even more problematic than it was before—which is no mean feat. The neat distinction between reports and instructions breaks down, and concepts such as *meaning* and *representation* reach new depths of murkiness.

But even after information has lost its innocence, it stays true to its original mission. Communication, honest or dishonest, can be instrumental in the creation and preservation of form—whether the form of the cellular slime mold, the form of a multi-cellular body, or the form of a corporation or government. Communication binds complex organic organizations, in defiance of the spirit of the second law. Information is the ingredient that evolution—whether genetic evolution or cultural evolution—adds to simple matter and energy to generate complexity.

Which leads to the next question.

CHAPTER

EIGHTEEN

WHAT IS COMPLEXITY?

Language makes it easy to put things off. "Tues: Cln bthrm" can be jotted down in no time at all, as can "Wed: Cln bthrm" and "Thurs: Cln bthrm." So it takes hardly any effort to jettison today's goals and reset them for tomorrow. If there were no short-cut to writing "Wednesday: Clean bathroom" and "Thursday: Clean bathroom," rescheduling would cost two to three seconds more each day than it does. That would add up. Eventually the more conscientious procrastinators might start supplementing their daily agendas with stern resolutions like "Qt prcrstntng!"—which, presumably, would also have to be spelled out in full, complicating things still further.

Underlying language's compressibility is its predictability. The *u* after *Q*, for example, does not come as much of a surprise, and can thus be deleted at no great informational cost.

Usually, compression is not so simple, because usually predictions cannot be made with such confidence; *e* is not the only letter to follow *Cl*, nor even the most likely. In a typical stretch of written English, only 20.5 percent of the words beginning with *cl* have *e* as their third letter. More likely to follow *cl* are *a*, which stands a 27.8 percent chance, and *o*, at 27.3. So *Clean* is far from the only possible result of adding vowels to "Cln." Still, by reading beyond "Cln," to "bthrm," we can expand the basis for prediction and thereby eliminate other possibilities. It is probably safe to say, for example, that the word *clone* seldom precedes the word *bathroom*.

Dropping vowels is one of the cruder ways to compress language. People who study the transmission of information have more sophisticated and more powerful techniques. They can cut a book in half without losing a word (although the result is pure gibberish to anyone without a decoder). This achievement will never be surpassed. It requires exploiting every iota of predictability in the English language, and thus leaves no room for further compression. In the technical literature, this fact is stated very compactly: English is 50 percent "redundant."

Redundancy (or predictability, or compressibility) is a two-edged sword—a procrastinator's best friend, perhaps, but a perennial irritant to the crossword puzzle designer. Having begun a vertical word with *q*, he is all but compelled to proceed with *u*, regardless of whether that fits into his horizontal plans. To a lesser extent, *cl* also narrows options. For the puzzle maker, the ideal language is a random one, in which all letters are equally likely to follow any given letter. He finds the subtle statistical order of English confining.

Things could be worse, though. English could be even more predictable than it is. The first person to point this out was Claude Shannon. His 1948 paper, "The Mathematical Theory of Communication," contained, in addition to a working definition of information and lots of arcana, some interesting observations about language. For instance, if English were much more predictable than it is—if its redundancy were, say, 80 percent—building enough crossword puzzles to sustain the game's popularity would be virtually impossible. But if the redundancy were only 33 percent, Shannon suggested, crossword puzzle designers would have so much leeway that three-dimensional puzzles might become the next recreational craze. And if it were 25 percent, four-dimensional puzzles would be practical (though unlikely, one gathers, to achieve commercial success). As things stand, the English language sits midway between redundancy—the predictability that allows procrastinators to waste time so efficiently—and randomness—the freedom of choice that, in greater abundance, would make puzzle building child's play.

All of this, it turns out, is relevant to the question of what life is and how it came into being.

Some people find it hard to believe that a heartless, brainless, spineless bacterium floating around in the primordial ooze could have evolved into a multibillion-celled animal that can agonize over lost loves, debate the nature-nurture question, and exceed a score of 10,000 in Pac-Man. This is a classic case of misplaced incredulity; the last 1.5 billion years of our evolution are not really that amazing.

Given a few prolifically reproducing organisms, occasional genetic mutations, limited quantities of food and territory, and lots of time, great evolutionary strides are all but inevitable. That is what makes natural selection one of the most appealing theories ever: great complexity follows from a few simple assumptions.

More remarkable than evolution itself is the prerequisite for it: the fact that the bacterium—or, even earlier, a bare, self-replicating strand of DNA—was floating around in the first place. That does not follow from anything. And, as we've seen, it goes against the grain of the second law of thermodynamics. While not forbidding the spontaneous formation of structure, the second law deems it highly unlikely.

Some scientists nonetheless attribute this seminal congregation of molecules to chance. Given enough time, they say, really unlikely things will come to pass—such as strands of DNA that make copies of themselves. But other scientists—Charles Bennett, for one—think that the first form of life owed its existence to some as-yet-undiscovered law of thermodynamics, one decidedly more upbeat than the second. This law would dictate that *some* systems, given certain advantages (access to energy, for instance), will grow in complexity, just as surely as most systems, under most circumstances, will not. This unformed law, says Bennett, has "taken over one of the jobs formerly assigned to God."

Among the obstacles to the law's precise formulation is confusion over the meaning of "complexity." People have lots of complexity; dogs have their share; worms have a little; bacteria still less. On that we would all agree. But can't we be more specific? How much more complexity do humans have than dogs? How much more does a bacterium have than the primordial soup? These are important questions. In physics, law isn't law until it takes mathematical form. A purportedly powerful equation wouldn't attract much respect if its numbers and symbols trailed off into the words "and then things get complex, sort of like a dog or a worm, but less so."

Bennett, working on the borderline between physics and computer science, shares the perspective that seems to come with that territory; he views just about everything as being reducible to information—in the physicist's sense of the word—and thus doesn't make a gross distinction between organic systems and other forms of information. People, dogs, pi, *War and Peace*—they're all conglomerations of information, as far as Bennett is concerned, and he's looking for a gauge of complexity that applies to them equally. If he succeeds, then, there will be doubly bad news for anyone who takes offense at this attempt to summarize human beings with a single number: he is placing us not only on the same scale as dogs, frogs, and slugs but also on the same scale as software, tax forms, and copies of *The National Enquirer.*

In pursuing a definition of complexity, Bennett begins with Shannon's discovery that information can be defined in the same terms as entropy. As we've seen, one way to phrase Shannon's observation is to say that ordered systems leave little uncertainty as to their contents: order has lots of information about its microscopic state embedded in its macroscopic description. An equivalent way of making the same point is to say that ordered systems admit to concise description. This is the best way to think of things in trying to understand Bennett's attempt to understand complexity.

For example, consider Bennett's first candidate for definition of complexity. Suppose, he says, that we define complexity as being the equivalent of order. This seems like a reasonable enough idea; if the second law of thermodynamics is an enemy of order, and life is in some sense at odds with the second law, then mustn't there be a close relationship between life and order—outright synonymity, perhaps? Some biologists talk as if this were true, as if order were indeed the essence of life's structure. It would be nice if they were right, because order, when identified with conciseness of description, is seen to be a very straightforward concept. Thus, the utmost in order is a string of letters like zzzzzzzzz, which can be described very concisely as $9 \times z$.

Unfortunately, no one has ever seen a person, or even a dog, that looks anything like zzzzzzzzz. A much more apt physical analogue of a string of Zs is dry ice. Its carbon dioxide molecules are arrayed so geometrically that a whole cubeful could be described with a statement like $CO_2 \times 10^9 \times 10^9 \times 10^9$. So if we defined complexity as order, we would be calling a smoldering block of ice more complex than a person. As Bennett has pointed out, this doesn't seem right. Order, then, is not the essence of complexity.

To be sure, order is *part* of anything that we would call complex. The toenail of a dog, for example, can be described economically: nothing but toenail molecules, as far as the eye can see. Indeed, order

is deposited in blobs and strings all over the dog. Still, the whole
dog is not orderly. It is that, here and there, but it isn't *just* that.

The word *complexity* itself suggests that we go to the other extreme
in trying to define it. By "Oh, it's so complex," people often mean
that something is indescribably complicated. So perhaps the most
complex things are those that most stubbornly *resist* economical description. Take, for example, a randomly produced sequence of letters—like bsohljgrijmtyhplrylvbd, which admits to no description
more concise than itself. Perhaps bsohljgrijmtyhplrylvbd is a shining
example of complexity.

Alas, this definition of complexity also runs into early trouble:
most dogs look nothing like bsohljgrijmtyhplrylvbd. Then again, you
may ask, what does? Well, if you assigned each of twenty-six inter-
mingled gases a letter of the alphabet, and then took an inventory of
molecules, your list might begin something like bsohljgrijmtyhplrylvbd. (This illustrates the fact that systems high in entropy—
such as randomly mixed gases—resist concise description.) Since dogs
are commonly considered more complex than gases, we must conclude that difficulty of description is not the hallmark of complexity.

Granted, *parts* of dogs are difficult to describe. There is a lot of
randomness coursing through canine veins: red and white corpuscles,
and other things that refuse to align themselves neatly. Still, the *whole
dog* doesn't look like bsohljgrijmtyhplrylvbd.

The whole dog looks more like "Tuesday: Clean bathroom"; it is
orderly enough, structurally predictable enough, redundant enough,
to be described with *some* economy—as, say, "Tues: Cln Bthrm." But
its description cannot be boiled down much further than that. Dogs,
and people, and bacteria, are like sentences, or paragraphs, or words;
they are a mixture of order and entropy, of redundancy and randomness. This is more than metaphor, says Bennett; there is an important
sense in which defining complexity is like answering the question
"What do words have more of than other arrays of letters?"

The obviously tempting answer is meaning, but this answer is
obviously problematic. Complexity, after all, resides in a physical
system even when the system isn't doing anything; presumably one
could, if armed with a good definition of complexity, assess the
complexity of a dog without watching it bark or fetch a stick. Yet it
would be hard, if not impossible, to gauge the meaning of a word

without seeing how it functions in everyday conversation and what behaviors it elicits in different contexts. Besides, no one knows how to quantify meaning. People don't even agree on what meaning means.

Bennett, though, who has meticulously examined and confidently rejected the various proposed definitions of complexity (not just those discussed above, but half a dozen or so others), thinks that quantifying meaning and quantifying physical complexity are essentially the same problem. Indeed, he has come up with a definition of complexity that, he believes, solves both problems. It is a definition so farfetched as to appear useless. And therein lies its significance.

To gauge the complexity of something, Bennett says, we must first find the most plausible theory of its origin. Since Bennett, like all good scientists, defines plausibility in terms of Ockham's Razor, this means we are looking for the simplest explanation of how the thing in question came to be.

The concept of the "simplest" theory, though a cornerstone of science, may sound too vague for a definition of complexity to hinge on. But Ray J. Solomonoff, now a scholar at large in Cambridge, Massachusetts, gave it some rigor in 1960, when he showed that choosing the most elegant explanation can, in principle, be reduced to counting bits. Competing theories, Solomonoff argued, could be translated into competing computer programs, each of which would generate a simulation of the phenomenon under study. If two programs yield simulations of equal accuracy, then the program that occupies the least memory space represents the winning theory. Bennett very much likes Solomonoff's idea and believes that it justifies including the phrase "most plausible theory" in his definition.

Once we have selected the most plausible theory of origin, the next step in gauging a thing's complexity is to reconstruct the process of creation this theory implies and count the number of "logical steps" along the way. In the case of human beings, the most plausible theory of origin is the theory of natural selection, and the number of logical steps is roughly the estimated number of times that our genetic material has been amended since it first began replicating, back in its prebacterial days. That number, says Bennett, is the measure of our complexity. Similarly, a book's complexity—the quantity of its meaning, in Bennett's terms—could be gauged by following the twists and turns in the sequence of events that most plausibly led to its being written, including the evolution of whatever sort of organism most plausibly wrote it. (Among the influences on Bennett's thinking about complexity was a short story by Jorge Luis Borges called "Pierre Menard, Author of Don Quixote," in which a writer attempts to recreate *Don Quixote* word for word without reading it—by reliving Cervantes's life as exactly as possible.)

Bennett concedes that his definition is a bit on the impractical side. In addition to its dozen or so obvious difficulties is the fact that the number of possible theories about anything is infinite. So how could we consider all possible explanations before picking the winner?

He has an all-purpose reply to such criticism: practicality is too much to ask of a definition of complexity. The most we can hope for, he says, is a definition that is an intellectual catalyst, one that helps scientists think about things more clearly and, perhaps, deduce the missing law of thermodynamics from other laws. When it comes time to actually *test* the law, his definition will be of no help; it cannot, in the real world, measure complexity. This, Bennett thinks, is not an indictment of his definition, but a tribute to complexity—to its intrinsic elusiveness, its subtlety, its, well, complexity.

It should come as no surprise, really, that this concept so doggedly resists analysis. Entropy itself, which seems elementary by comparison, is in fact very hard to pin down; it is easy enough to *say* that systems high in entropy are those that most stubbornly resist economical description, but in practice this idea suffers from a difficulty much like the one that afflicts Bennett's definition of complexity: we can never be sure that all possible descriptions have been considered. One could stare at the number 0.2857142 for days without realizing that it can be expressed succinctly as $2/7$. Arrays of letters, and of molecules, can similarly conceal their order.

The ideas of entropy and complexity have something else in common besides—and perhaps related to—their subtlety: they open our eyes to subtle distinctions. There was a time when people assumed that all forms of energy were essentially the same. Then came the concept of entropy, which highlighted the difference between energy that is so disordered, so dissipated, as to be useless, and energy with enough structure to be useful. Thus, the benign first law of thermodynamics, which guarantees that the amount of energy in the universe can never change, surrendered center stage to the unsettling second law: the amount of *energy* may stay constant, but the amount of *useful* energy declines continually, as entropy accumulates. The idea of complexity, a shade subtler than the idea of entropy, is giving our picture of the universe higher resolution still, by highlighting the distinction between two kinds of entropy; systems can be equally disordered yet differ in complexity. The other side of this coin is that systems can be equally ordered yet differ in complexity. Thus, there is trivial order, such as a cube of dry ice, and nontrivial order, such as, we humbly submit, portions of ourselves. Once again, a new distinction hints at a new generalization: even as the amount of order in the universe decreases, something called complexity appears to be growing.

But will we ever know by how much? Will the day come when we can confidently say that the average American is 2 percent more complex than the typical Neanderthal man or 98 percent as complex as the Holy Bible? One glance at the junk heap of would-be definitions of complexity suggests not. And Bennett's definition may also lead to that conclusion—not only if it, too, fails, but even if it succeeds. Great complexity, by this measure, is beyond measure; to make his gauge work, Bennett has made it unworkable. Perhaps that is a necessary compromise, so long as the items being evaluated can do anything from clean bathrooms to create crossword puzzles—and, on any given day, can choose to do neither.

PART V

KENNETH BOULDING

CHAPTER

NINETEEN

SEMINAR AT THE SAGE

The fact that Kenneth Boulding is a Quaker does not mean that he looks like the Quaker on the cartons of Quaker Oats. However, as it turns out, there is a certain resemblance. Both men have shoulder-length, snow-white hair, blue eyes, and ruddy cheeks, and both have fundamentally sunny dispositions, smiling much or all of the time, respectively. There are differences, to be sure. Boulding's hair is not as cottony as the Oats Quaker's, and it falls less down and more back, skirting the tops of his ears along the way. And Boulding's face is not soft and generic. His nose is jutting, and his eyes are deeply set and profoundly knowing.

Boulding is sitting at the head of a long assemblage of tables, wearing a navy blue blazer, a baby-blue V-necked sweater, a pale blue Oxford shirt, and a cowboy tie—silver-tipped, braided leather strings dangling from a turquoise stone. He can often be found in this getup, or some slight variation on it; traveling light is important in his line of work. Ever since being forced into retirement at age seventy, he has been in the visiting professor's "racket," as he calls it, traveling from college to university to think tank, teaching and studying wherever people will pay him to do so. This spring he is at the Russell Sage Foundation, in a sleek, black, six-story glass-and-steel structure that stands anomalously amid turn-of-the-century town houses on Manhattan's Upper East Side. He is now in the basement lunchroom, where he is about to be the centerpiece of a ritual. Today is his seminar day. He is going to talk about the project that has been occupying him for these past few months—and, off and on, for the past twenty years. The title of his talk is "The Logic of Love."

Most of the people around the composite table are visiting scholars, scholars in residence, or administrators. At a nearby table are miscellaneous support personnel, and they have the look of veterans.

This is not the first time they have eaten the ham-and-cheese sand-
wiches, nibbled at the salad, then sat back, donned inquisitive expressions, and endured. Nonetheless, this will be a new experience
for all of them. No matter how many seminars they have sat through,
they have never been witness to anything quite like what is about to
happen.

As basements go, this is an elegant room. It is centered around a
large, square, ground-level skylight, and the walls are appointed with
large, horizontally rectangular examples of what might be called
Kleenex art, for two reasons: (1) it could well have been created by
taking wadded-up Kleenexes, dipping them in pastel paints, and
dabbing them on canvas; and (2) the resulting pattern is so soft and
inoffensive that it could be used on decorator boxes of Kleenex. In
fact, it looks so familiar that I'm not sure it hasn't been.

The middle-aged woman sharing the end of the table with Boulding is a sociologist. She taps on her glass with a spoon and, after the
chatter subsides, asks everyone to look their best; a photographer is
roaming the room, and the pictures may find their way onto a bro-
chure. "Look beautiful for posterity," she says with a laugh. "Look
alert." Boulding laughs, too. This is something he does a lot. He
laughs generously at other people's jokes and unabashedly at his own,
which come forth with monologic frequency, often in the service of
exposition or argument. He illustrates the Heisenberg Uncertainty
Principle, which says that the act of measurement affects the thing
measured, by reference to a recovering heart patient who, when asked
how he's doing, says "Fine" and is killed by the effort. He expresses
his reservations about the concept of economic equilibrium by calling
it "useful"—pause—"useful for passing examinations with." When
asked to describe his niche in academia, he says that he is "an idea
man and court jester."

The sociologist gets things rolling. "Now I'd like to formally introduce Kenneth Boulding, who has been a legendary figure to me
ever since my awakening to social consciousness and social movements. He counted among the voices of authority and reason that the
people of my world listened to on matters of concern to humanity. I
knew he was an economist, and that gave his voice further legitimacy.
I didn't know he was a poet." It's true. He has written a book of
religious poetry called *There Is a Spirit: The Naylor Sonnets*, published

by an obscure press in Nyack, New York, as well as a pamphlet of poetry—whose publication he underwrote—in praise of his wife, called *Sonnets for Elise*. Sometimes when academic conferences get more ponderous than usual, he writes limericks about the participants to lighten things up.

"Ken Boulding is international both in background and in outlook," the sociologist continues. "He studied at Oxford and briefly was a fellow at the University of Chicago. He has taught at many universities in England, the United States, and Japan. A sample of his titles: distinguished professor of economics emeritus, University of Colorado; past president of the American Association for the Advancement of Science; past president of the American Economic Association; past president of the Peace Research Society and of the Society for General Systems Research." A camera clicks and flashes, and film advances with a whine. "Prizes and awards include thirty honorary degrees, which are reasonable recognitions for his contributions as an economist."

Boulding interjects, "I never took a Ph.D., you see," and then laughs. Boulding speaks in an accent that, to the British ear, betrays his roots in working-class Liverpool but, to my ear, just sounds British and therefore authoritative. Sometimes he stutters, though not as badly as he did as a child and young man. When the affliction strikes, it is as if he were trying repeatedly to crush an ice cube with his wisdom teeth, and when it subsides, it is as if he has succeeded. The stutter makes Boulding even more likable than he might be otherwise. It removes any trace of snobbery from his accent and thereby accentuates his geniality—which hardly needs it; he is the most obliging and agreeable man I have ever met; he uses the word *yes* the way children use ketchup. One time I was interviewing him, with my tape recorder running, when his phone rang. His end of the conversation went like this:

Mr. Boulding speaking. Speaking, yes. Oh, hello, hi, how are you? Yes. Yes. Yes indeed. In fact, it's almost done. I just have to get it from my secretary in Boulder, and you ought to be getting it at least in a couple of weeks. That's right, yes. Right, yes. Oh, that's too bad, yes. What a pity. Oh, I think so. I think so. Uh, let's see, now let me just get this, let's look at my diary here. This is June twentieth? Yes, I'd be very glad to do that, yes. Right. Right. Just introduce it and have a discussion. Yes. Excellent, yes. Very good, yes. Just fine, yes. As I say, you'll certainly get it in a couple of weeks, yes. Right, yes. Yes. Excellent. Okay, fine. Look forward to seeing you there very much. Yes, bye.

The only problem with Boulding's unrelenting graciousness is that during serious discussions it is hard to tell whether he agrees with you or is just being nice. "Yes, of course, precisely" means, in this context, "Either I don't understand what you've said or I understand what you've said and either agree, disagree, or have no opinion."

The sociologist continues, "I hesitate to report to my colleagues with a tendency to writer's block that Professor Boulding has written some forty books." This, too, is true. Not all have been smashing successes, but none has suffered from lack of ambition. They have titles such as *Human Betterment*, *The Meaning of the Twentieth Century*, *Conflict and Defense: A General Theory*, *A Reconstruction of Economics*, and *Beyond Economics: Essays on Society, Religion and Ethics*. The sociologist says, "We've already been enriched by our casual conversations with Professor Boulding, and by occasional thoughtful papers that have provided us with intellectual treats in the last few months. And we now look forward to hearing his perspective on the logic of love."

Or so she thinks. "Well, I'll get to love later on," Boulding says right off the bat. What he wants to talk about first is his most recent book, *The World as a Total System*. In it he questioned the usefulness of looking at the world as a mosaic of national systems—national cultures, national economies, national communications networks. Increasingly, he wrote, national economies are intertwined with one another so thoroughly as to constitute a unified, worldwide economic system. More generally, the dense web of international communications channels has brought the planet to the brink of becoming a "single social system."

Boulding begins, "Let's look at the earth from outer space, now—that wonderful, beautiful, blue-and-white globe that we've all seen pictures of." Such a view, he says, "wouldn't tell you much about the social system, I'm afraid, because it's mostly under clouds." So imagine, instead, a miniature representation of the earth, a globe. A globe is "an extraordinary condensation of information"—political boundaries, population centers, geographic contours. But even so, it, too, offers scanty insight into the social system. So, "suppose we made it a globe on which you could see this building," he says. "That would be a thousand feet in diameter—about the size of the Empire State Building. . . . If you wanted to see individual human beings on it, you'd have to get it larger than that. Maybe a mile in diameter, I don't know. Anyway, that's what we're looking at. We're looking at four and three-quarters billion human beings in the space throughout the globe. And, uh, we're looking at human artifacts; that's part of the social system. And we're looking at the noosphere, as Teilhard de Chardin calls it—that is, what's in here"—he points to his head—"all the knowledge and so forth and content of the minds of these four and three-quarter billion humans."

It's probably a mistake to try to say what is going through the minds of the people in the Russell Sage dining room at this point. Being in the middle of a seminar, they are all more or less poker-faced. Still, I get the feeling that there is a certain amount of puzzlement floating around the room; it isn't clear what, if anything, Kenneth Boulding is driving at.

"And then we might want to put some invisibles on the globe—trade routes, something of that sort," he continues. "And then, of course, we have to recognize that it isn't just a globe, it's a movie of a globe. We have to put it in four dimensions as well as three, so we have the space-time dimension of it, which is very much like a movie, as we think of it, I think—just a succession of frames, one after the other. We go from one to the next and all sorts of things have happened."

The puzzlement seems more tangible now. Indeed, the vibrations emanating from some corners of the room speak of outright alarm. This is not surprising; the seminar is nearly an hour from over. In Manhattan, the only people who have time for fifty minutes of free association are psychiatrists, and they get paid for it.

"And it's process, isn't it—as opposed to structure?" No one volunteers an answer. "So we certainly have to look at process, and this involves things like aging and death and birth. While we're sitting at this table, maybe ten to twenty thousand people died in the world and considerably more than that have been born—that is, if they do it on the hour, at the hourly rate." Somehow, this fact leads within a few sentences to the subject of mathematics and its relation to reality. "One of my big songs and dances, really, is that the real world consists of shapes and sizes and structures and is very fundamentally topological," he says. He adds, "I've sometimes argued that there are probably only four numbers in the real world: e, pi, 0, and 1. You can derive all the others from that. Even the velocity of light isn't a number, after all." He laughs at this. Few other people do.

Before the hour is over, Boulding will discuss the concept of power; the great mystery of how people change identities so readily (he, for example, has changed from an Englishman, a Methodist, and a chemist to an American, a Quaker, and an economist); the equally puzzling phenomenon of legitimacy (What, he asks, is the difference between a tax collector and a mugger? The tax collector has legitimacy); human sacrifice among the Aztecs; hatred as a cybernetic phenomenon; the Vietnam war; the rising divorce rate; the decline of American agriculture; and other things.

To the extent that Boulding's talk has a unifying theme, it is the same theme that unifies—to the extent that it does—another of his recent books: *Ecodynamics*. It is a theme that is classic Boulding— eclectic, amorphous, immense, and aimed ultimately at saving the world. Human society, according to Boulding, is held intact—and, sometimes, torn apart—by three great "systems," three great categories of human relationships: the "threat system," which depends on coercion and is embodied in policemen, bank robbers, playground bullies, and armies; the "exchange system," which depends on reciprocation and is embodied in shoppers, merchants, bankers, and baseball card traders; and the "integrative system," which depends on—well, that's the problem. Boulding is still trying to get a grip on the integrative system, and he can't yet characterize it with satisfactory precision. It has to do with churches and charities and brotherly love and families. But it has to do with a lot of other things, too: duty, legitimacy, patriotism, fanatical devotion to the Boston Red Sox—so many things, indeed, that sometimes it isn't clear what the integrative system *doesn't* have to do with. Boulding confessed to me once, "I've never really been able to reduce integrative systems to a system." He has come to Russell Sage in hopes of coming as close to that reduction as possible, and finally finishing his book on the integrative system, the book that is to be called *The Logic of Love*.

The longer Boulding talks, the more difficult I find it to look the other scholars in the eye. It isn't so much that *I'm* embarrassed for him. I've read enough of his work to know that beneath this crazy quilt of insights and opaque quips is a reasonably coherent body of thought. It's just that I suspect that *they* are embarrassed for him. To the uninitiated, after all, there appears to be no real purpose in his ramblings. I'm afraid that if my gaze ventures beyond the region of space between Boulding and the surface of the table, I will see the embarrassment in their faces. Then they will see that I see the embarrassment, and there will be all sorts of awkwardness floating around. Finally, though, curiosity gets the better of me, and I glance around the room. I am pleased to see that the scholars are being good soldiers—eyes fixed solemnly on Boulding or on the table. No one is balancing a checkbook or clipping fingernails. A few people even have pens poised, ready to record the notes that will inspire their stimulating questions. As far as I can tell, though, no one—with the exception of the sociologist—has actually written anything down. She has written:

threat
integrative
exchange (money)

The hour is almost up before it dawns on me that most of these people are confused as to the source of their embarrassment. They think they are embarrassed for a seventy-six-year-old man whose powers of concentration are in decline and who therefore jumps aimlessly from topic to topic, covering his tracks with moderately funny one-liners. This is wrong. It's true that he is seventy-six years old, and it may be true that his concentration span is beginning to contract, but the fact is that Kenneth Boulding has always been a disconcertingly discursive thinker. You can go back and read things he wrote ten, twenty, thirty years ago and you'll find the same exasperating pattern: you follow a thought for a page or two, and then, just as you are getting a feel for it, it is lost in another thought that will soon suffer the same fate. It seems, sometimes, as if everything Boulding says is an aside. His asides range from grand principles (his summary of world history is "Wealth creates power, and power destroys wealth") to grandly frivolous metaphors (Hitler and

Stalin were "pimples on the changing countenance of time"). At the
end of one of his books, one always has the sense that things of great
import have been said, but it would be difficult, in many cases, to
summarize them without using nearly as many words as the book
contains. After laying down *The Meaning of the Twentieth Century*, and
relishing the considerable new knowledge it had imparted, I had only
one question: What is the meaning of the twentieth century?

Part of the frustration of reading a Boulding book originates in his
method of operation. He dictates his books, working sometimes without even an outline, and often with little revision; he has been known
to complete one in less than a month. This pace leaves no time for
meticulousness, and his writing, like his thinking, is notable more
for its boldness and intermittent brilliance than for its organization
and consistent correctness. Among his mottos is "Don't get it right—
get it written."

Another part of the frustration originates in the architecture of
Boulding's mind—in his tendency to wander freely from one discipline to the next and to think, as one colleague put it, "on several
different levels of organization at once." He illuminates the transition
from peace to war by reference to water that is being cooled and then
frozen. He writes that Ohm's Law, though "exhibited in its purest
form in the study of electricity," is also applicable to the flow of
money. He has soberly analyzed the similarities between frogs, fishes,
and bacteria (on the one hand) and Baptist churches, post offices, gas
stations, families, counties, and states (on the other). He once told
me, "The automobile's a species just like the horse; it just has a more
complicated sex life." Gossip, he says, is like DNA, only more prone
to mutation upon replication. He has used equations of epidemiology
to study the spread of hostility and friendliness, and has drawn
diagrams that look for all the world like something out of an economics textbook, except that the two extremes on the horizontal axis are
labeled "virtue" and "vice."

The odd thing about Boulding's free-floating, in some ways fragmented, thought is that it grows out of a quest for unity. In fact, it
grows out of a particular kind of quest for unity; it is not merely an
intellectual eccentricity of Boulding's—though it is certainly that—but
is somewhat characteristic, more broadly, of something called general
system theory, of which Boulding was one of the founding fathers.

General system theory is not in vogue now. It was born in the 1950s amid exaggerated promise, and it declined under the weight of excessive expectations, among other burdens. But it has its merit as a way of looking at things, and it has its relevance to the subject at hand. Like reductionism, general system theory harbors as one of its ideals the unification of the sciences. And, like Edward O. Wilson's reductionism, it underscores the significance of information at different levels of organization. Indeed, it depicts the world as a hierarchy of information-processing systems, somewhat in the spirit of Edward Fredkin. It attaches so much significance to the concept of information, in fact, that in Europe general system theory is synonymous with cybernetics.

But general system theory and reductionism are, in important ways, not at all alike. Their methodologies differ, and their underlying philosophies stand in opposition; representatives of the two schools tend to disagree on the big, old issues, such as determinism versus free will. And when it comes to the question of the meaning of information—and the meaning of the information age, and, for that matter, the meaning of life—general system theory (at least, Boulding's brand of it) says more suggestive, more cosmic, and markedly weirder things than does Wilson's reductionism.

CHAPTER

TWENTY

THE BOULDING PAPERS

Kenneth Boulding's cultural information resides, among other places, in the Bentley Historical Library, on the North Campus of the University of Michigan. It is kept in file folders that are kept in cardboard boxes that are one foot wide and fifteen inches long. There are forty-six boxes. A few of them house mainly things written by Boulding's mother or wife, and scattered through most of the boxes is the work of other people—letters from friends and colleagues, selected articles relevant to his work. But most of the papers are his—letters, manuscripts, published articles, notebooks, doodles, sketches, poems, lengthy typewritten accounts of his travels. The Boulding papers end in the mid-sixties; the last twenty years of his writing—between fifteen and twenty boxes, assuming a constant rate of output—has not yet been deposited. The librarian keeps asking him to send them along from Boulder, Colorado, where the Bouldings make their permanent residence, and Boulding keeps saying he'll get around to it soon.

The Bentley Library was built in 1972 and is in all respects clean. Its structure is unadorned, and its carpet and tables and long stretches of thick glass wall are spotless. It has no smell. But if you open one of the boxes that hold the Boulding Papers, lean over it, and inhale deeply, you may briefly imagine that you are in the most remote recess of the oldest library in the world. The box containing his undergraduate lecture notes—recorded in small, cloth-bound notebooks with thick, stiff, yellow pages whose edges, like those in an old family Bible, form a marbled pattern—is enough to make you sneeze.

Boulding was prolific from very near the beginning. His first surviving poem came circa age seven.

We have such a dear little kitten
her coat is all shiny black
And she's just like a little fur mitten
our dear little tiny black cat

For a time, at ages seven and eight, he kept a diary. ("February 24, 1918. The Doctor came tonight and just once had the impudeinse [*sic*] to put a spoon down my throat.") At age eleven he put out a weekly neighborhood newspaper, using the family's Oliver typewriter and carbon paper. ("Editors Remarks: This is a new paper, indeed, the thought of it only originated in my mind a few days ago, but I hope it will prosper, till, from being a mere childish whim, it will become a paper taken by everyone in the street.") When children were mean to Kenneth, he gave them unfavorable coverage, a practice that quickly elevated his status and influence. By his late teens, Kenneth was an accomplished letter writer. He typed his letters and kept carbon copies. In the upper right-hand corner of each letter he typed "4 Seymour Street, Liverpool." There he had been born on January 18, 1910.

Liverpool was then, as it is now, a working-class city with pockets of extreme poverty, a gritty and grimy place to grow up. (Boulding's standard line about Liverpool: "I was taught in school that trees turned green in the spring, and I thought that meant they got moss on their trunks.") Kenneth suffered no sense of deprivation, though. His family was at least as well off as the neighbors, and his neighborhood sparkled by comparison with the Irish slums. And, being an only child, he received plenty of attention.

Boulding's parents were squeaky clean—the kind of people whose moral judgments a marginally sinful neighbor lives in fear of. They did not drink or swear. Kenneth's first taste of alcohol would come in middle age, after he read that moderate drinkers are healthier than teetotalers. Kenneth's father, Will, was by vocation a gas fitter and plumber, and by avocation an urban missionary and volunteer social worker. He preached in the slums, took poor children out to the country on weekends, and at times opened his home to the homeless. Mrs. Boulding—Bessie—assisted in this benevolence, worshipped devoutly, and wrote religious poetry. At the bottom of letters written to Will before their marriage, she might add, as a shorthand post-script, "Psalm 23rd."

While Kenneth was away at Methodist summer camp or, later, at college, Bessie was sometimes moved to write letters to him in verse.

> (My pen has gone to Derbyshire
> and so I cannot write,
> Except in pencilled letters
> to the son of my delight!!
> etc.)

Kenneth inherited both his mother's attraction to verse and her tendency to get mushy. Upon mailing her a mock newspaper he had produced while at camp, he enclosed a note that justified itself in these terms:

> This is just to get in the only thing you can't get in a newspaper—Love. Give as much as you can to everybody and keep the rest yourself.
>   With Tons and Tons and Tons of it,
>   Ken

This was when Kenneth was sixteen.

Bessie had made a deal with God before Kenneth's birth: if given a boy, she would do her best to make him a missionary. She didn't press Kenneth overly hard on fulfilling this promise, he stresses; she was not maniacally pious; she had a sense of humor. Still, it would have been hard, in that family, to escape feelings of religious obligation. By age ten, Kenneth had read the Bible, Genesis through Revelations (in part to secure the one-shilling reward his grandfather had offered). On his grandmother's wall was a piece of religious imagery that has stayed clearly in his memory. It depicted two roads: a broad road, lined with pubs packed with hedonists, leading straight to hell; and a narrow, tortuous road, along which Christians trudged, leading up a mountain, through the clouds, and into heaven. Behind and above the scene was an immense eye, the eye of Judgment.

All of this notwithstanding, Boulding did not uncritically embrace his parents' religion. The Methodist Church supported the national war effort, and the war's impact on Kenneth was severe. It is hard

for a boy of six or seven to assimilate news about cousins and neighbors who will never come home and whose parents will never be
quite the same. The first entry in Kenneth's diary—January 1, 1918—
is "1,252 DAY OF THE WAR." Every day thereafter is so numbered,
and from August through early November these headings are the
only entries. The November 11 heading is "PEACE WITH GER-
MANY," and on that day his regular writing resumed.

At age fourteen, Kenneth suddenly became a pacifist. "I had this
experience in which I felt that if I was going to love Jesus, I couldn't
participate in war. I mean, the conflict, you see, between the teach-
ings of the Sermon on the Mount, especially, and the phenomenon
of war seemed to me so overwhelming that I couldn't really do both."
A few years later, a friend invited him to the Liverpool Society of
Friends meetinghouse for Sunday worship. The experience was
strange and powerful. There was no altar, no clergy, no laity—only
people, all with equal access to God. When the meeting began, silence
enveloped the room—a silence "as impressive as the finest speech,"
Kenneth wrote to his aunt that evening. The silence lasted for an
hour, punctuated only by a few spontaneous utterances that were
"theoretically under the direction of the Holy Spirit," as he now puts
it. "The very first meeting I went to, I felt at home. I felt this was
my spiritual home, which it still is."

Kenneth was not to earn prestige on the playground. He was un-
coordinated and a little sickly, and the idea of sports struck him as
ludicrous. But his intelligence soon came to the attention of teachers.
He was naturally interested in astronomy, and when the school dis-
trict's inspectors came around to check on educational efficiency, he
was trotted out to recite and explain Kepler's laws. The headmaster
at the Hope Street School selected him and a friend for special
tutoring, and at age twelve Kenneth won the Earl of Sefton Scholarship to Liverpool Collegiate School, the only conduit through which
a working-class boy in Liverpool stood even a remote chance of
reaching Oxford or Cambridge. Teachers prepared students assiduously for scholarship examinations and tried also to prepare them for
the world beyond, instilling the manners, even the accent, of the
English gentry.

The record of Kenneth's late adolescence is in letters to his child-less—and, one gathers, lonely—Aunt Ada. From age seventeen through graduation from college, he sat down every Sunday night, with few exceptions, and wrote detailed and whimsical narratives of the week's events. He routinely produced 1,000 words, sometimes 2,500. The image of Boulding that emerges from these letters is that of a teenage boy who could profit from an occasional dose of valium; his exuberance, while clearly genuine, is relentless. Not atypical greetings: "Dear Auntie Ada"—"O Aunt!"—"O! My Aunt!"—"My Sainted Aunt!" Not atypical description of dinner: "We had chicken for dinner today. My, it was nice!" Not atypical opening stretch of letter:

Dear Auntie Ada:
Of Poets that bloom in the Spring, tra-la
   Tra-la-la-la-la-la-la-LA
O come let us joyfully sing, tra-la
While verse is so free on the wing, tra-la,
   Tra-la-la-la-la-LA
   For Daffies are blooming,
   And so I'm assuming,
We're soon to be favoured with Spring

In March of 1928, Kenneth wrote Aunt Ada about the family's reaction to opening the *Times* of London and reading, under University News, "Elected to an open Major Scholarship in Natural Sciences at New College, Oxford, Mr. K. E. Boulding, of Liverpool Collegiate." "Then the world went round. Mother wept and went pink all over and Dad and I went oogly inside. Then I dashed off to school. I have nearly had my hand shaken off today."

College proved a rude awakening. Boulding had long been aware of class distinctions, but they had never been driven home with such force as at Oxford. His Liverpool accent had not been fully subdued at Collegiate, and the "public school boys" (as private school boys are for some reason called in England) were not eager to associate with a boy from the lower classes—least of all one who stuttered profusely. This is not to say he spent his years at Oxford alone; he had a faithful group of friends "among the outcasts" and occasionally was even befriended by a member of the "lower gentry." But he was never allowed to forget where he had come from. Years later he wrote to a friend, "My first term at New College left a scar which nothing afterwards quite effaced, and in spite of the good time I had afterwards when I had gathered unto myself a circle of friends of more or less like disposition I don't know that I have ever quite recovered."

Kenneth's scholarship to Oxford was in science, and he matriculated as a chemistry major, but from the beginning science had to compete with the humanities for his attention. He had literary aspirations, and they were not his alone. One year, after he failed to win the Newdigate Prize for poetry, his mother wrote:

My Dearest Dear:
Drat that Miss Fielding, I should like to smack her good and hard! Carrying off that prize that I wanted you to get so badly!

From a later prospect, not winning the Newdigate Prize would look like a blessing. After the winner for Kenneth's freshman year was announced, he reacted to the loss with characteristic wholesomeness—by taking a twenty-mile bicycle ride, during which he decided to forsake chemistry and follow his heart; he switched his major to "modern greats" in the School of Politics, Philosophy, and Economics. Becoming an economist, he reasoned, would enable him to serve humanity. The two most serious problems in the world, he had felt for some time, were war and unemployment, and in 1929 the latter seemed the more pressing.

When Boulding entered Oxford, he was a socialist, and there he joined the Labour Club. But the allure of Marxism did not survive the experience of reading Marx. (Boulding recites the apothegm: "A man who isn't a socialist before he's twenty-five has no heart, and one who is after he's twenty-five has no head.") Meanwhile, John Maynard Keynes was formulating a theory that seemed to surpass Marx's in explanatory power. Keynes tied cycles of economic growth and decline, then widely considered inevitable, to patterns of government spending and taxation. When Boulding read Keynes's *Treatise on Money* in 1931, all of economic history since the Roman Empire fell into place; "this came as a blinding kind of revelation to me," he says. He told Cynthia Earl Kerman, who worked as his secretary

before writing her Ph.D. dissertation about him and publishing it as
the biography *Creative Tension*, "You see, this was a feeling that *now*
the world made sense. . . . Here was a *mind* really at work, who was
a *much* greater man than Marx. . . . This was a—I would say it was
a spiritual experience as well as an intellectual one."

Boulding graduated in 1931 as the top-ranking student in economics and then stayed at Oxford for a year of graduate work. The
following spring, on the brink of having to assume worldly respon-
sibilities, he won a reprieve—a Commonwealth Fellowship, kind of
a Rhodes Scholarship in reverse, which provided a $6,000 stipend,
plus various allowances, for two years' study in the United States.
He went to the University of Chicago to study economics.

After recovering from his first encounter with New York—"noisy,
untidy, and smelly, and overcrowded, and vulgar, and for the most
part, desolatingly ugly," he wrote to a friend—Boulding fell in love
with America. Though the American intellectual class was a "hopeless minority," he found it "more open minded, less prejudiced, and
much less pompous than our educated classes." As he puts it now,
"Nobody asked who your grandfather was." And what class consciousness there was worked to his advantage; his hybrid dialect of
English, though still too plebian to hide his heritage at home, had a
patrician air in the American Midwest. (His status as a distinguished
foreigner and an Oxford honors student, he would later surmise,
partly explained his ability to secure teaching jobs in America without
a Ph.D.) Suddenly, Kenneth Boulding was a valuable social commodity. From Chicago he wrote to a friend, "My time is divided
between eating enormous quantities of food, having violent argu-
ments, economic and theological, going to innumerable parties given
by innumerable charming young ladies, and generally enjoying myself according to the best Oxford standards."

England did not let go easily. While Boulding was in America, his
father died, leaving his mother in severe financial straits. Will had
been such a lenient creditor, and had done work so cheaply for
churches and other worthy customers, that his liabilities now outweighed his assets, 1,248 pounds to 860. Kenneth returned to England in 1934 and, after a summer of job hunting, landed a teaching
job at the University of Edinburgh for 250 pounds a year. He and
his mother moved to Scotland, and he settled in for three years of

work in an environment that he found intellectually and socially
barren by comparison with Chicago.

Boulding's letters from the mid-1930s show his two sides to be
developing rapidly. First, there are the letters of the young scholar
in search of advancement. He writes humbly and literately to the
eminent economists of the day, and the letters bear fruit. Friedrich
von Hayek sends Boulding favorable critiques of his articles; Joseph
Schumpeter, under whom Boulding briefly studied during a summer
at Harvard, writes a testimonial letter calling him "one of the most
promising economists of his generation . . . well on the way of acquiring an international reputation." The editor of Harvard's *Quarterly
Journal of Economics* compliments Boulding on his articles, asks for
more, and exclaims, in closing, "My dear boy, you seem to have
ideas." (News of the first acceptance of a Boulding manuscript had
come years earlier, when he was twenty-one, in a letter from John
Maynard Keynes himself, then editor of *The Economic Journal*.)

Even as Boulding's academic stature rose, his correspondence reflected also the continued growth of the religiously motivated social
activist. In 1935, he typed up "A Plan for the Establishment of an
Order of World Citizenship" and circulated it among friends. Though
the plan today reads like a parody of youthful idealism, it was not
radically out of tune with the times. Boulding had come of intellectual
age in the wake of World War I, a time of much serious speculation
that world federalism was the only route to stable peace. Among the
most forceful proponents of this view was H. G. Wells, whose *Outline
of History* had made a deep impression on Boulding when he read it
as a teenager. Now, amid the first faint stirrings of another world
war, Boulding decided to take action. "If our world is to survive,
national sovereignty must go," he wrote. "As political organisms
Britain, Germany, Japan, and all the rest of these petty states must
suffer the fate of Wessex and Mercia; must be absorbed in a greater
whole."

Boulding realized that the odds against this goal were high, but,
being twenty-five years old and divinely inspired, he was undaunted.
He wrote: "On the side of nationalism are ranged all the powers of
poetry and song, language and literature, self-love and pride; in its

service are gathered the strongest emotions, worthy as well as base; loyalty and love as well as lust and hate. On the other side we have only the cold light of the human intelligence, warmed by the small fire of complete unselfishness. But on this side is truth, and on this side is God, and either it will prevail or we shall go down into utter darkness." Boulding's plan—laid out in a section subtitled "The Vision"—was to organize a small, but always growing, body of people who would cease to identify with their homelands, relinquish the entitlements of national citizenship, and establish an "interpenetrating state organisation of their own" that would provide education and social services, even administer criminal justice. But this was still a ways off. "The physical organisation must wait upon the growth of mental body, the spirit of world consciousness." The first step toward this spirit was to draw up a document called "An Affirmation of World Citizenship," whose signatories would pledge not to fight in a war and not to accept the military protection of any nation. They would be known the world around by the sign they would affix to their signatures—a cross within a circle: the symbol of unity.

The friends of Kenneth who reviewed his plan to save the world were not overwhelmed by its practicality. One wrote back that "people just aren't made that way, to be able to cooperate enough to obtain agreement on any given desideratum and put it into effect. . . . If you want to see a universal world political power established, much the more effective way of establishing it is by pushing imperialist plans of one bunch or the other, until they extend their dominion over everything." In the face of this and other such reactions, Kenneth was embarrassed but unbowed. He conceded that his plan had been "stupid and pompous" and assured his friend that it had "died stillborn." He wrote, "I see now that I made the mistake of trying to erect what was for me an intensely personal experience into something of a universal dogma." Still, he would continue to pursue the cause, to consider himself a citizen of the world. "I think you underestimate the importance of spiritual forces in bringing about political unity, and the necessity of spiritual forces for the maintenance of unity; it is only so long as the mass of the people *believe* themselves to be 'Americans' that 'America' can exist as a stable political unit."

While in Edinburgh, Boulding applied twice, unsuccessfully, for fellowships at Oxford, and the rejections opened old wounds; he was convinced that his social class had worked against him. He wrote to a friend, in a truly rare display of malice: "Sometimes I hate Oxford and all it stands for more than I hate anything else." In the midst of this refreshed alienation, he had what Quakers call "a leading"—an illumination of one's future that, in terms of vividness, lies somewhere between an intuition and a hallucination. It happened while he and some friends were climbing Mount Snowdon, the highest peak in Wales, without climbing gear. "We came down a rather dangerous route, really," he recalls, "down a practically vertical cliff. I got a bit scared. I was just a mountain scrambler. But still we kept going, and I had this kind of feeling that I was going to go back over to America. It practically just came out of nowhere." Kerman describes Boulding's emigration in less transcendental terms: "Nothing succeeds like failure; he had failed as an Englishman, so he became an American."

His excuse to return to the United States came in the summer of 1937 in the form of the Friends World Conference. His excuse to stay there came via a friend from his Chicago days who was then teaching at Colgate. Colgate, it so happened, had just lost an economic theorist to the Roosevelt administration. Would Boulding be willing to take his place? The $2,000 salary was tempting, but it wasn't enough to support Kenneth and his mother, so he declined. The offer was upped to $2,400. "Every man has his price," Boulding notes. He sent for his belongings and, a year later, for his mother.

On May 4, 1941, Boulding was in Syracuse for a Quarterly Meeting of the Society of Friends. Halfway through the meeting, he was moved to speak. "I hate moonlight," he began. It was moonlight, after all, that illuminated the targets for the bombers then decimating Europe, and it was the moon that so well symbolized the romantic illusions that sustain war. The moon was thus a half-light, shining on the half-truths and false emotions, emotions that would not stand up to the light of day. Boulding implored the Friends to dwell in the whole light of God. "As I sat down," he wrote a few months later, "I caught the glance—half amused, half rather awed—of a tall, golden-haired, sunny-skinned and blue-eyed girl sitting a row or two directly in front of me." She was Elise Biorn-Hansen, a Norwegian-American in her early twenties who, it so happened, had been deeply influenced by her mother's pacifism and had decided, after hearing

Norman Thomas lecture, that she would go through life as a citizen of the world.

On May 7, Kenneth began a letter to Elise "Dear Elise." On June 5 it was "Dearest Elise." Then: "Dear Heart," "My Love," "Totally Beloved." On August 13: "Heart of Mine." On August 31, Kenneth and Elise were married. They moved to Princeton, New Jersey, where Kenneth went to work for the League of Nations, analyzing problems in European agriculture.

CHAPTER
TWENTY-ONE
AMBUSH AT THE EAGLE TAVERN

The Eagle Tavern looks at first glance like any other New York working-class bar: stools; booths; large, surly bartender; Yankee fans who will grudgingly watch the Mets on the TV; Met fans who will grudgingly watch the Yankees. But beyond this room, through a double door, is another room, much larger and a little less usual. It is the kind of place where an Elks Club might hold its annual dinner/ finance meeting. The walls are veneered with dark wood, and along them hang large stuffed and painted fishes—a tarpon, a sailfish, a blue marlin caught off Key West. A vividly blue shark hangs just beyond my booth, which is upholstered in red leatherette. Lest the marine motif be lost on anyone, the doors have signs that say things like MATE'S QUARTERS and GALLEY, and there are a couple of round mirrors with porthole frames. Even the light fixtures on the wall— fake candles encased in brass-embellished glass—are vaguely sugges- tive of a nineteenth-century sailing ship. Overhead, inexplicably, the light fixtures are miniature wagon wheels, suspended from the acous- tical tile ceiling with chains.

Up front—where, if this *were* the night of the Elks Club dinner/ finance meeting, a man would be tapping a water glass with a spoon— sits Philip Boulding, tuning his harp. He picks up a magic marker and paints the C strings red. The F strings are already blue. Philip is a tall man, not much past thirty, with long arms and broad wrists. He has naturally wavy hair, parted down the middle and combed back over his ears, somewhat like his father's. He also has his father's eyes, or at least his father's *look*. It is kind of an innocent leer, a leer with nothing sinister behind it. It is the way Dracula would look at you if he were a nice guy.

A few people mill around the room as an almost random array of notes emanates from Philip's harp and from the nearby hammered dulcimer being played by his wife, Pam. The two of them constitute Magical Strings. At the back of the room, through the double door, Kenneth appears, slightly stooped, smiling his odd smile. He pays his six dollars without protest. This is a gross miscarriage of justice, not just because he is Philip's father but also because he has lately been serving as Philip's public relations agent. He sent a memo around the Russell Sage Foundation announcing this one-night stand, and he spread the word at Penington, the Quaker boardinghouse where he is living this spring. No telling how many of the eighty or so people who will soon occupy this room can be credited to Kenneth.

He walks toward his son, leers at him lovingly from twenty feet, and says, "Hi, Philip." Philip looks up, leers back, and says softly, "Hi, Dad." Kenneth says, "Your sign's fallen down." Indeed, the banner hung by the sponsoring organization is folded over, obscuring itself. When righted it will read PINEWOODS FOLK MUSIC CLUB in crazy-quilt fashion, each letter cut from a different cloth.

Now it's time for ambush journalism. Kenneth doesn't know that I'm here, and I don't know him well enough to be sure he'll even recognize me; I've interviewed him once at the Russell Sage Foundation, more than a month ago, and once at George Mason University, his previous patron, about six months ago. I've been trying to call him for a week now to set up our encounter tonight, but he's been out of town. I'm beginning to wonder if we'll even have an encounter; after he said hello to his son and began walking back toward the rear of the room, I made eye contact with him and tried to nod the way people who know each other nod, but it must have looked like the kind of nod exchanged by polite strangers, because he just returned it and kept walking. He is with two men, and they are finding a seat toward the back of the room. This is bad news.

But, just as my despair is entering its preliminary phase, Pam walks back and retrieves him, escorts him to the table right next to mine, and introduces him to a woman who is seated there. They have met before, years ago, and Pam reminds him of this. He is standing right next to me, exchanging pleasantries with the woman. The pleasantries wind down. This is my chance. "Professor Boulding," I say. He turns and looks at me. I introduce myself and remind him of our conversation a month ago. "Oh, yes, well, of course," he says. "Of course, yes." Then, with no prompting, he does just what I want him to do: he sits down at my booth, right next to me.

I'm sorry to surprise him like this, I say—I tried getting in touch with him. He explains that he's been in Stockholm, at an inaugural meeting of some international institute for the social sciences. The weather was so bleak, the discussion so abstract, that he finally resorted to writing limericks to amuse himself. For example, a paper called "The Decline of the Superstates; The Rise of a New World Order?" he boiled down to this:

> It may not be too bad a sign
> When Superstates start to decline,
>   For what they are Super
>   at may be a blooper,
> To which we should never incline

Boulding is happy to report that the conference featured scholars from nearly a hundred nations, including two from China and a number from Africa. He is sad to report that the Arab nations failed to send a single representative. It has been only nine days since the American air strike against Libya, and all talk about the Middle East still leads there. Boulding, as is his wont, has formulated a one-line capsulization of his opinion on the issue: "You don't deal with terrorism by becoming a terrorist," he says.

Boulding's stream of consciousness now flows in short order through discussion of (a) what an odd fellow Muammar al-Qaddafi is; (b) how politically unwise it was for him to try to change the Moslem calendar; and (c) the parallels between the Moslem and other faiths. "I've often said there are three Judaic religions: Christianity, Islam, and Marxism." He laughs at the joke, which I only dimly understand. But seriously: the major religions do have similar injunctions, such as the Golden Rule. And, Boulding adds, "they all have some concept of sin."

Talk turns to Philip. "He makes all those instruments," says his father, pointing proudly up front, where two Celtic harps sit by two hammered dulcimers. "He's remarkable, really. He's completely self-taught. He didn't go to college, but anything he wants to know, he'll find a way to learn." The irony, says Kenneth, is that, although much of Magical Strings' repertoire is Irish (mainly songs written by the blind Irish bard Turlough O'Carolan), neither Philip nor Pam has a drop of Irish blood. "That's a great argument against sociobiology," Kenneth laughs. On the other hand, memes don't appear to go much further than genes toward explaining Philip's talent. Elise plays the cello a bit, and Kenneth the recorder, but they didn't force either on their son. "Our philosophy of child rearing was creative neglect," he says. Then he rears his head back and gives me his leer, laughing.

The five Boulding children entered the world with rhythmic precision, one every two years from 1947 to 1955. ("We just got in the habit of having a child ever other year and it took us ten years to break it.") Creative neglect was not all-encompassing; when the subject turned to morality, Kenneth and Elise tried to instill the "Quaker values"—essentially, the idea "of looking at the human race as the children of one father."

Kenneth's oldest child is now a geologist in Indiana who lives off the land, testing soils for pay and, along with his wife, growing vegetables. He wrote an historical novel about the conversion of Iceland to Christianity and started a publishing house to publish it. The second oldest child is a conceptual artist whose works include a photographic study of skid marks and a pit dug on the outskirts of Boulder, filled with distinct layers of clay and earth, and hidden from public view by the grass growing on top. ("But he knows it's there. And then in three million years it will erode and people will see it, if there's anyone there to see it.") Two other Boulding children lead more conventional lives: the daughter is a banker—or was, until her second child was born—and the youngest son teaches marketing at Duke.

And then there is Philip, the fourth of the five. His birth convinced Elise that she would never have time to earn her cherished doctorate in sociology, so she and Kenneth decided to turn him into a surrogate Ph.D. with the name Philip Daniel. Philip doesn't share Kenneth's opinion that little of his musical interest and talent was inherited from his parents. He remembers falling asleep to the sound of Kenneth playing selections from Bach on the piano. (Kenneth had failed to mention not only Bach but the fact that he plays the piano.) When

Philip was six, his mother asked him what instrument he wanted to play. Philip chose the violin and stuck with it for years, but upon entering adolescence he took up guitar. He began building guitars in the Bouldings' basement and soon decided to make a career of it. The idea of becoming an instrument maker, not especially practical to begin with, fell further into question after Philip, at age seventeen, got hooked on hammered dulcimers. A friend introduced him to them—not to the dulcimers themselves, but to some recorded dulcimer music and a book about them. Though Philip had seen dulcimers only in pictures, he built one and learned to play it. Then he built another. Then another. For four years he built dulcimers that were the only ones he had ever seen, guided by what he heard on records and by his inner sense of tone. Even today, he says proudly, no dulcimer sounds quite like a Philip Boulding dulcimer.

All of this Philip's parents endured stoically. They did not talk much about the shrinking market for hammered dulcimers, or about spending more time on homework and less in the basement, or about the importance of a college education. They seemed to understand, says Philip, that he would have to find his own way through this. Besides, they had little time to counsel. Elise finally had earned her Ph.D., and the family had moved from Michigan to Colorado after the University of Colorado offered both parents teaching jobs.

Though Kenneth worked very hard, it was not high-pressure hard. Philip didn't feel like the son of a corporate lawyer, or even of a harried academic; Kenneth simply was often absorbed—blissfully, it seemed—in his work. That he had little time to spend with the children individually (family outings were fairly common) seems only to have intensified Philip's admiration. He was always in awe of his father, aware of his renown and therefore all the more impressed with his genuine humility, his nearly constant joviality, and his ability to put anyone at ease. (The Bouldings' house was a place where Kenneth's students and colleagues felt at home, and a nerve center for various protests during the war in Vietnam; Tom Hayden, for example, was an occasional guest.) In the lives of many young men comes a time when they realize that their father is not, as they had innately assumed, omnipotent or infallible, that he is just another man, with strengths and weaknesses. Philip doesn't recall such a time. "I still sort of view him as some sort of deity," he says.

Now, an ominous development: the woman with whom Kenneth was exchanging pleasantries before I hijacked him has decided to join us. She crosses the aisle, clutching her bottle of Guinness, and plants herself on the other side of the table, noting, by way of explanation, that her table was getting a little crowded. She is a very large woman, and a very colorful one. She is wearing pink earrings, pink fingernail polish, pinkish lipstick, and a pink pullover top, mercifully subdued by a black sweater, unbuttoned, and a black winter scarf. (It's April, but the temperature is in the thirties, and snow is expected tonight.) Pinned to her sweater is a button that says MY LIFE NEEDS EDITING. I don't doubt it, but I don't especially want the job, especially when I'm trying to steer a discursive seventy-six-year-old man toward a discussion of the meaning of life. And yet, I fear, this may be my fate tonight.

Sure enough, it's all on the table in no time flat. She's divorced—he was a painter, a creative man, a romantic man, but, well, you know the kind—and now she's sort of drifting around, following, at the moment, Magical Strings on their east coast tour. She wants to write a book about Pam, a friend of hers from way back. This project was revealed to her in a dream. "It's something I should have known a long time ago, and it was just shown to me," she says. She loves Pam's children, and one of her highest aspirations is "to be a grandmother without being a mother." I look at her oddly. "It's possible," she insists, noting that she's already a godmother. This remark triggers a series of observations by Boulding that culminate in this one: "Being a grandfather is like being a visiting professor. You have all the fun and none of the responsibility."

Now the large woman starts asking me what I do for a living. Do I write fiction? Do I read poetry? The truth comes in handy here: no and no. I explain to her that I work at a magazine called *The Sciences*. No sooner are the words out of my mouth than I am seized by a brilliant idea: it so happens that the most recent issue of the magazine is in my briefcase, and it occurs to me that it just might serve as a pacifier; she shows no signs of scientific inclination, but the magazine is illustrated with fine art, and if I could get her leafing through it, maybe Kenneth and I could resume our discussion, after which politeness would keep her from interrupting. I reach down, pull the magazine out, and set it in front of her on the table.

Then something goes terribly wrong. Boulding picks it up. "Yes, oh, lovely," he says. "Yes."

He begins flipping through the magazine, and before long it is apparent that the printed word has a mesmerizing effect on him. He proceeds rapidly and without distraction, scanning some paragraphs and reading others closely. He happens upon a long review of a book called *The Dynamics of Apocalypse: A Systems Simulation of the Classic Maya Collapse* and begins reading it steadfastly. About every great person you could ask the simplistic but nonetheless illuminating question: What is the source of the greatness? Perhaps this is the answer in Kenneth Boulding's case: he can read a mile a minute under any circumstances.

He is oblivious to the woman standing at the microphone, asking that people refrain from smoking, and to her introductory words of praise for Magical Strings. Finally, when Philip and Pam walk out on stage to applause, he looks up and claps. Then, before his son even sits down, he returns to his reading, still clapping. What I would give for this kind of concentration.

The music begins—dreamy music, strings softly plucked or lightly hammered—and Boulding keeps plowing through the magazine, pausing now to absorb an advertisement for a series of books from Columbia University Press whose scope happens to coincide with his realm of expertise: FROM THE MOLECULAR TO THE MAMMOTH, the blurb reads. His son, in the midst of "Downfall of Paris," pulls out a "penny whistle" and begins to play it. This instrument is his most direct musical legacy from his father; Kenneth bought one at age nineteen, while at Oxford, and it led him to the recorder. Kenneth looks up as soon as he hears the sound. He gazes at his son for half a dozen seconds and returns to reading.

Philip and Pam are quite a team. Pam plays the dulcimer from her soul, banging her hammers as if in a trance, swaying to the music with mindless contentment. Philip is more detached, more analytical; he watches the strings as he plays, willing his fingers to pluck and glide, pluck and glide. The two of them make, well, beautiful music together. When they first met—he was teaching hammered dulcimer, and she was in his class—both were married. The attraction was mutual and strong, and the result was "like an Italian opera," according to Kenneth.

Kenneth is sinking ever deeper into his reading. He taps his feet
to the music occasionally or thrums his fingers, but by and large he
is absent. So is Philip. He has gotten carried away, like Pam. His
eyes are now closed, and his fingers are drawn as if by magnets to
their intricately arranged destinations. Philip was the problem child;
he was so late to talk and to read, and so awkward in his early
movement, that at first his parents suspected brain damage. He hated
high school, hated college, ran off and got married at eighteen. He
is the only one of the five children to have divorced. But now everything is all right, and he and his father are in the same room, immersed in their own worlds, both having heard different drummers.

The song is over. Philip nods with grateful humility to the applause. He and Pam begin to play an Irish jig, and Kenneth continues
to turn the pages. Finally, he closes the magazine, takes off his reading
glasses, puts them in his pocket, and rubs his eyes. Forty-eight hours
ago he was in Stockholm, where it is now 4:00 A.M., according to
the internationally calibrated watch on his wrist. He rests his left
elbow on the table and places his face in his hand. His eyes narrow.
His head moves up and down so slightly that I'm not sure whether
he's keeping time to the music or merely breathing. He stirs and
repositions himself, placing both hands in his lap, crossing his legs,
and letting his head droop. The music stops. Applause. Boulding
opens one eye and claps half inversely, his right palm against the
back of his left hand. He stirs again and folds his hands on the table.
He is snoozing unabashedly now, head hanging down, a long shock
of gray hair falling over his forehead, nearly touching the tip of his
nose.

There are two dangers here. One is that Boulding will have a
snatch of nightmare or a sudden muscle contraction and unconsciously launch the empty bottle of Miller Lite now positioned on
the thin stretch of table to the right of his right hand. The other
danger is that in some other, less spectacular but equally embarrassing
way it will be revealed to the crowd that the father and father-in-law
of the stars of the show is asleep. This latter possibility grows markedly as Philip talks about the dream that inspired him to write
"Astronomer's Dream," which he and Pam are about to play. In the
dream, he journeyed across outer space to visit a planet, a beautiful
planet, tranquil and green—"much like earth must have been before

human habitation." And though it wasn't clear why he had come to this planet, it was clear that there was a reason; there was a purpose; his journey had meaning. Now, as his father slumbers, Philip induces what I alone am in a position to see is a crisis of potentially great magnitude: he announces, "I'd like to dedicate this song to my father. I feel lucky if I see him once a year, and he's here tonight." If it were within my power, I would freeze time and resuscitate Kenneth and then melt it.

No need. Kenneth's head is up before the *r* in father, and he looks as alert as at noon. Philip continues, "He has dedicated his life to peace and peace research, and I like to imagine that the planet I saw in my vision might be a place we could create right here on earth." The people clap, and Kenneth faces them and smiles with warm grace. He sticks his right hand up in the air and waves awkwardly— arm straight and vertical, hand equally straight; all the bending takes place at the wrist. He is as adorable as a little child.

CHAPTER

TWENTY-TWO

THE ORGANIZATIONAL
REVOLUTION

World War II was a trying time for Kenneth Boulding's pacifism. It was difficult to believe that a just God would oppose armed resistance against Adolf Hitler. In a poem called "Out of Blackness," Boulding later recalled his reaction to Germany's invasion of Belgium and Holland in May of 1940:

I feel hate rising in my throat
Nay—on a flood of hate I float,
My mooring lost, my anchor gone,
I cannot steer by star or sun;

Later that month, though, he had an epiphany, an apprehension of the unity of humankind. He was drying himself off after a bath when "I had this experience of, of the crucifixion, essentially," he recalls. Heading off any misinterpretation, he adds quickly, "It wasn't a hallucination. It was just a very strong internal experience, you see, of being there." He saw that Jesus had died for everyone, Briton, American, and German alike. His hatred "just kind of ran out of me and went down to the floor with the water. I just felt completely drained." This enlightenment was recounted in the climactic lines of "Out of Blackness":

Hatred and sorrow murder me.
But out of blackness, bright I see
Our Blessed Lord upon his cross
His mouth moves wanly, wry with loss
Of blood and being, pity-drained.
Between the thieves alone he reigned:
(Was this one I, and that one you?)
"If I forgive, will ye not too?"

Boulding's newfound tolerance for the Allies' enemies, however eloquent in expression, did not win wide favor. In the spring of 1942, about the time of the Bataan Death March, he and Elise drafted a statement calling on all nations to disarm. Even some Quakers found it hard to support, and Kenneth's boss at the League of Nations deemed it unprofessional. He gave Boulding a choice: either issue the statement as planned or continue to work at the League of Nations. He issued the statement. Boulding says that every time he has tried to make a noble career sacrifice, he has been rewarded with a better job. This time it was at Fisk University, a black college in Nashville that gave him a tenured position in the economics department.

After a year at Fisk, Boulding accepted an associate professorship at Iowa State College in Ames, Iowa. He went to Ames planning to study the economics of labor unions, a subject that drew him quickly into the study of various other aspects of human society. "It became perfectly clear to me that if you wanted to study the labor movement, you had to be a sociologist and an anthropologist and a political scientist," he says. "So I got interested in the unification of the social sciences."

He moved one more time before pursuing that interest in earnest. In 1949, when the University of Michigan was wooing him with a tenured position, Boulding was in a good bargaining position. He was happy at Ames, and he had just been selected to receive the John Bates Clark Medal, given every two years by the American Economic Association to the outstanding American economist under age forty. He said he would take the job only if he could teach a seminar on the integration of the social sciences. Each year's seminar would be aimed at assembling a general, interdisciplinary theory of some kind—a theory of competition and cooperation, say, or of information and communication. He got the job and the seminar, which promptly taught him that "the social sciences didn't want to be integrated very much." Still, with dogged recruiting, he usually succeeded in gathering a group of people who might have gone the rest of their lives without meeting had it not been for him. In 1953, when the seminar sought a general theory of growth, the participants included a physicist, a bacteriologist, a zoologist, a geneticist, an architect, a botanist, a linguist, a psychologist, a sociologist, an economist, and something called an economic zoologist. Boulding observes earnestly, "Once you start integrating anything, there's no place to stop."

Boulding's refusal to respect disciplinary boundaries gave him something in common with Ludwig von Bertalanffy, an Austrian-born scientist who was turning such disrespect into a discipline all its own. Bertalanffy, a biologist by training, had been struck by similarities of pattern among different levels of organization. Equations from statistical mechanics, developed to describe the mass behavior of molecules, can be applied to traffic flow, or to the diffusion of rumors. And, as Norbert Wiener had explained, some patterns of information processing, notably feedback, appear in organic systems ranging from the cellular to the social. Bertalanffy was a man of large vision and ambition, and he decided the time was ripe to codify laws governing systems in general (human beings, human societies, corporations, frogs, ponds, oceans, etc.) or some subset of those systems (say, corporations and frogs). General system theory, as he called the enterprise, was an attempt to integrate the sciences that, unlike some past attempts, did not rest on reductionism. Principles of psychology and sociology would not *follow* from principles of the "hard" sciences at lower levels of organization; they would parallel them. The different disciplines would be drawn together, like so many friends, because they have so much in common. "Unity of Science is granted," Bertalanffy wrote, "not by a utopian reduction of all sciences to physics and chemistry, but by the structural uniformities of the different levels of reality."

The rejection of reductionism was not just a by-product of general system theory but part of its impetus; Bertalanffy had a pronounced philosophical agenda. He had long been a critic of the prevailing, "mechanistic" approach to biology, insisting that not all things are best studied through analysis of their components. Some biological properties, he said, can be understood only by looking at the organism as a whole. This line of thought, which Bertalanffy called the "organismic" approach, is a form of holism, the general belief that wholes are more than the sums of their parts. From the beginning, this approach carried some of the philosophical and even ideological baggage that in recent years has led people to associate holism with vegetarianism, Zen, and communes. Bertalanffy wrote:

The world is, as Aldous Huxley once put it, like a Neapolitan ice cream cake where the levels—the physical, the biological, the social and the moral universe—represent the chocolate, strawberry, and vanilla layers. We cannot reduce strawberry to chocolate—the most we can say is that possibly in the last resort, all is vanilla, all mind or spirit. The unifying principle is that we find organization at all levels. The mechanistic world view, taking the play of physical particles as ultimate reality, found its expression in a civilization which glorifies physical technology that has led eventually to the catastrophes of our time. Possibly the model of the world as a great organization can help to reinforce the sense of reverence for the living which we have almost lost in the last sanguinary decades of human history.

Boulding's view of the sciences was not so explicitly political in its slant, but he, too, had doubts about reductionism. In 1954, he expressed them with a more concrete metaphor:

There is a famous hotel in Eureka Springs, Arkansas, which leans up against a hillside in such a way that any one of its floors can be entered from the ground. The hotel of knowledge is something like this: each of its floors can be entered from the ground; we do not have to know everything about physics and chemistry before we can move on to biology, or everything about biology before we can move on to the social sciences. Each floor can be explored as an independent unit and by its own methods, and the elevator service goes down as well as up. Biologists can learn things from the social scientists, and physicists from the biologists.

In 1953, Boulding wrote Bertalanffy a letter that Bertalanffy later quoted in his book, *General System Theory*:

I seem to have come to much the same conclusion as you have reached, though approaching it from the direction of economics and the social sciences rather than from biology—that there is a body of what I have been calling "general empirical theory," or "general system theory" in your excellent terminology, which is of wide applicability in many different disciplines. I am sure there are many people all over the world who have come to essentially the same position that we have, but we are widely scattered and do not know each other, so difficult is it to cross the boundaries of the disciplines.

In the fall of 1954, Boulding and Bertalanffy found themselves together in Palo Alto, California, as visiting scholars at the Center for Advanced Study in the Behavioral Sciences, then in its inaugural year. Bertalanffy had been conducting an interdisciplinary seminar at the University of Chicago on a general theory of behavior, and two of the seminar's regulars had also been invited to the center: Ralph Gerard, a physiologist, and Anatol Rapoport, who then was a mathematical biologist and later became known for his work in game theory. One day while the four men were sitting around a lunch table, their impulses to integrate found synergy. They were discussing the obstacle to enlightenment posed by disciplinary boundaries when, Boulding recalls, an almost tangible unity of purpose dawned. "We just sat down at that lunch table and"—he snaps his fingers twice—"it went like that. It became *so clear* what we ought to do." What they did was start a society devoted to general system theory. "We drew up a manifesto around the table and put it into *Science* magazine." Noting that the four of them "had come to this from completely different directions," Boulding adds, "that was an integrative experience, I'd say." Thus was born the Society for General System Theory, soon renamed the Society for General Systems Research. The first meeting was held at the next convention of the American Association for the Advancement of Science, and seventy people showed up. Boulding was elected the society's first president.

Boulding's departure from mainstream economics had been a while in the making. In 1941, Harper and Brothers (now Harper and Row) had published his first book, *Economic Analysis*. It was a college text, and a perfectly respectable one—idiosyncratic in its free use of wit and metaphor, perhaps, but not radical in conception. It found favor with mainstream economists and went through four editions before being eclipsed by Paul Samuelson's *Economics*. Boulding's next book, *Economics of Peace*, was a bit offbeat—written, as it was, during World War II. But it was published in 1945, so its prescription for postwar recovery turned out to be well timed, and the book found a solid audience in several nations. Boulding's third book, *Reconstruction of Economics*, published in 1950, was not just unconventional but an attack on convention—a proposal to overhaul the tools of economic analysis. Most of his colleagues either rejected or ignored the argument, and thereafter his professional identity moved away from economics and toward social theory at large. His next book, published in 1953, was neither economics nor, strictly speaking, social science, but a work of science (social and otherwise) and ethical philosophy. *The Organizational Revolution* was also one of the first books written within the framework of general system theory.

It was so general, in fact, and so eclectic, that a reader could be forgiven for wondering what, if anything, fell outside the book's domain, and where, if anywhere, the line lay between serious analysis and fanciful metaphor. Boulding wrote, for example:

> The coexistence of different ecosystems can be explained by quite slight changes in the critical variables as we move around the geographical or social landscape. Thus in the Midwest certain slight differences in temperature and humidity set the geographical boundary between prairie and forest. In society likewise many widely differing social and ideological ecosystems coexist in the same region: the lush forest of Roman Catholicism and the open prairie of Boston Unitarianism, for instance, existing side by side in different "microclimates" or social strata of the same society.

Whatever the merits of this particular instance of free-form analysis, Boulding did put the general system perspective to profitable use in pursuing one of the book's main purposes: tracing the origins of "the organizational revolution"—figuring out why "the past fifty or a hundred years have seen a remarkable growth in the number, size, and power of organizations of many kinds," ranging from labor unions to corporations to governments.

Addressing this question involved a search for principles common to organizations of different scales. One example is "the principle of increasingly unfavorable internal structure," which states that, as an organized system grows larger, the concomitant demands on its structure and infrastructure may grow disproportionately. If you took a flea and somehow enlarged its dimensions until it reached the size of a dog, its legs would break under its own weight. A dog would suffer the same fate if inflated to the size of an elephant. The problem in both cases is that the animal's volume, and thus its weight, is growing cubically as the thickness of its supporting limbs grows linearly.

But such constraints are not etched in stone; they can be pushed back through technological innovation. Conceivably, for example, dogs' legs could be made of some organic alloy that would permit them to withstand elephantine pressure.

Similarly—and more relevantly—the limits imposed on organic growth by problems of internal communication can be pushed back through technological innovation. Thus, human beings could not have reached their present size had natural selection not discovered electrical information transmission; if our brains communicated with our extremities via microscopic mule train instead of rapid neural firings, we would not be able to walk down the street *or* chew gum, much less do both at the same time; we would be too big for our own good.

It was in comparable technological innovations, Boulding argued, that the origins of the organizational revolution lay. Beginning in the 1800s, the railroad and steamship stretched the scope within which materials could be economically transported, and the telegraph and telephone then made the instantaneous transmission of information possible over any earthly distance. "One has only to try to picture the large organization of today, whether it be a corporation, a government, a labor union, or a farm organization, operating without telephones and conducting all its communication through horse-drawn mails to realize the extent to which the telephone has contributed to the growth of organizations," Boulding wrote in *The Organizational Revolution*. "One might list also the improvement in the other mechanical aids to the recording, communication, and interpretation of information: the typewriter, the duplicator, the business machine, and finally, of course, the electronic calculator, which may have an impact on the structure of human organization beyond even its own calculation."

In a sense, the organizational growth that Boulding perceived was not news—not even century-old news. The scope of social organization has been expanding at least since people first formed villages, which, after all, then evolved into towns and cities. But even if we go back well before the organizational revolution, Boulding's basic thesis—his link between the state of information technology and the scope and complexity of social organizations—fares well. It is hard to imagine large cities, for instance, that lack the ability to preserve information for long periods of time. Financial records are a prerequisite for the large-scale taxation associated with urban life, and they greatly facilitate trade and thus the division of economic labor, which is also characteristic of even the earliest large cities. It is no coincidence, then, that these cities arose in the societies that first encoded information systematically. Indeed, the correlation between social and symbolic complexity is strikingly exact. The first great cities in Mesopotamia took shape between 4000 and 3000 B.C., just as a system of enumeration employing clay tokens was beginning to evolve into cuneiform. Memphis, similarly, emerged in Egypt shortly before 3000 B.C., the time of the earliest surviving hieroglyphic tablets. In the New World, Mayan cities evolved in tandem with a variant of hieroglyphics, while the Incas, who possessed only a relatively crude system of enumeration, inhabited villages that fell just short of truly urban complexity.

The growth of the state, no less than the advent of the city, had to await the appropriate information technology. The Roman Empire only reached its unprecedented compass with the aid of a system of roads that, given swift messengers, permitted long-distance communication at then-revolutionary velocity. And the content of this communication, in order to effect political control, had to be fairly complex; the administration of a territory so large would have been nearly impossible without a powerful and flexible system of symbols. Later, the printing press, the postal system, and, finally, various forms of instantaneous information transmission all had their effects on the nature and scale of human societies. It seems unlikely, for instance, that the economic and political fibers of the United States would have assumed continental cohesion in the nineteenth century without telegraphy. Human history, if viewed from a sufficient distance, can be seen as an erratic but relentless growth in the scope and complexity of organization, often on the heels of new information technologies.

None of this really explains *why* human society has grown more complex; it explains why it *has been able* to grow more complex. With *The Organizational Revolution*, Boulding had shown that social organizations can express their expansionary impulses only with suitable information technology; but he hadn't gone into detail about the source of those impulses. His contribution, in that regard, is somewhat like E. O. Wilson's discovery of pheromones. In both cases, the information technology that keeps complex organizations glued together is a separate issue from the impetus behind the evolution of that complexity. In the case of human society, the philosophical stakes of this second issue are high.

CHAPTER
TWENTY-THREE
DINNER AT THE PENINGTON

Kenneth Boulding's latest temporary residence is Penington House, a boarding home that since 1897 has been run by Friends for Friends. It stands on East Fifteenth Street, sandwiched between other old homes, just around the corner from the simple but stately red brick meetinghouse where worship services are held every Sunday. I am in the dining room, having my first encounter ever with a group of Friends. "Dining room," really, is misleadingly formal. It is a combination kitchen–den–dining room, and it is in the basement. Diners serve themselves from aluminum pie pans arrayed along a counter. The pans contain things such as spinach quiche, tofu casserole, and something that is labeled enchiladas but, strictly speaking, is not. This food appears not to be of recent origin, but the banana cream pie is first rate. No frills in the beverage department: milk and water, coffee and tea.

The dozen or so diners, seated at a long table that could hold many more, have nothing conspicuous in common, except their manifest harmlessness. (This is one of the few times since moving to New York two years ago that I've been in a group of any size and considered myself the most menacing person present.) Almost everyone appears to be under thirty-five. The only elderly resident, besides Boulding, is a woman named Mary. She has lived at Penington House for twenty-three years, longer than anyone else, and she loves it, in general; her only complaint is that it has no elevators; she has arthritis in her hips. Mary, who talks nasally and breathes loudly, is seated to my left, and Boulding is directly across from me, wearing his blue, blue, blue, and turquoise outfit.

I can only imagine what it is like for these people to have Kenneth Boulding in their midst for these few months. There is no exaggeration in saying that he is a legendary figure within the Society of

Friends. The nature of its worship places a premium on impromptu eloquence, which he has in abundance. During his years at Colgate, he was regarded by some Quakers as a prophet. (He plays down this prophet business.) His book of religious poetry and his pamphlets on peace can be found in many Quaker homes, including this one. His eminence within the distant world of academia adds to his aura.

Boulding is regaling us with observations about everyday life. This has been his first opportunity to live in New York, and he loves its ethnic and socioeconomic diversity and its sheer human drama ("It's almost as good as the theater"). He bemoans "a certain lack of maintenance" in parks and other public places, and the absence of *real* supermarkets—the wide-aisled kind, like King Sooper back home in Colorado. In general, though, Manhattan so completely meets his needs that he has ventured beyond it hardly at all since arriving here in February. His only excursion has been to Princeton, New Jersey, where he worked more than half a lifetime ago. He recalls that his League of Nations office was in the Institute for Advanced Study, immediately above Albert Einstein's office. "I always say," he says, "I could feel the emanations through the floor. I never had the nerve to introduce myself. I've always regretted this."

Einstein. His image appears in my mind, and I look at Boulding. Hmmmm. There's a certain resemblance: Boulding's hair, when in its more unkept phases, fans out a little like Einstein's; frontal silhouettes of the two men would be very similar.

"Was he the inspiration for your hair?" I ask. Ridicule is not a common pastime around Penington, and even such lighthearted needling as this is no laughing matter. A woman says earnestly to Kenneth, in his defense, "I think your hair is an expression of *you*." Boulding, alone among the Friends, gets a laugh out of my line, which reminds him of a true story. Each year, Boulder is the site of a huge Halloween party, and last year he was walking around amid the 10,000 costumed partiers when he ran into a threesome done up in a Wizard of Oz motif. "Here was the cowardly lion and the tin woodsman and—what was the other one?"

"The scarecrow," Mary says.

"Yes, the scarecrow. And he embraced me as the Wizard of Oz. But I wasn't dressed up at all."

It hits me for the first time: Kenneth Boulding looks remarkably like the Wizard of Oz—not the imposing, dragonlike wizard that spewed smoke and fire, but the little man at the controls behind the curtain. The resemblance seems so striking that I come within a hair of exclaiming "My God. The resemblance is striking!" But I think better of it; if a comparison of Boulding to Einstein was not well received in these quarters, there's not much favor to be gained by pursuing this line of thought. Instead I ask who played that role. No one knows for sure. Ray Bolger? someone asks. It wasn't a major role, Mary says. But it wasn't minor, I reply; the same actor played the phony fortune-teller Dorothy encountered in Kansas before returning home in time for the storm.

It occurs to me what an alarming image I've happened upon: Kenneth Boulding as charlatan. Is that what he is? A mush-minded do-gooder who finessed his way into the academic limelight with his wit and British charm? Is all this stuff about the integrative system just a Christian Trojan horse, a sermon about peace and love dressed up in scientific terminology and wheeled into the citadel of secular humanism?

The conversation turns to Boulding's fifth book, *The Image*, in which he wrote about the various ways that images—internal representations of the outside world—influence behavior and suggested that perhaps it was time to start a discipline devoted to images, called eiconics. The book was well received, especially among students of mass communication. But, thirty years later, the birth of eiconics has yet to take place. I was reading *The Image* on the subway tonight and was struck by a point Boulding makes in it: it is easy enough to say, as is sometimes said, that purposeful behavior is the hallmark of living systems, but the fact is that some systems commonly thought of as inanimate could loosely be said to possess purpose. If you mix eighty hydrogen atoms and forty oxygen atoms under the right circumstances, they will form molecules of H_2O, as if it were their "purpose" to combine in predetermined ratios. And if you remove them from this "desired" state, they will "seek" a return to it—and, conditions permitting, succeed. Granted, the atoms do all this with no more intelligence than a padlock or its key; an atom just keeps bouncing off ill-fitting atoms until it finds one it fits snugly. But remember: it is just such lock-and-key mechanisms that stand between life and chaos; both our DNA and the cells it builds defy the spirit of the second law through information processing that boils down to molecules either fitting one another or not fitting.

I tell Boulding how interesting I found this point, and he says, "Oh, yes, well, I've always argued that it isn't just an accident that 'valency' comes from the word for value." Then, passing from the subtle to the opaque, he makes a joke about why CH_3 is known as a radical. And then, before I know it, he is talking about his nine-level hierarchical categorization of reality. It ranges from static structures, such as rocks, to "clockwork structures"—such as the planets, which revolve mindlessly, with no apparent feedback, no need for information—to things that do process information, such as thermostats, to things such as cells, which process both information and raw matter and energy, to plants, to animals, to human beings, to social systems. And then, at the very top, are "transcendental systems." I look at him askance when he gets to this level, and he laughs defensively.

"This is a hierarchy of evolution, in a sense," he says. What he means is that the direction of this hierarchy—toward greater complexity, more richly processed information, more elaborately pursued purpose—roughly corresponds with the direction that evolution has often taken. Not only was the organizational revolution just another phase in the long, slow rise in human social complexity; the long, slow rise in human social complexity was just another phase in the long, slow rise of organic complexity. In the beginning, complex molecules formed cells. Then cells got together and formed organisms. Then organisms got together and formed societies. And now, under the influence of new information technologies, human societies are approaching the intricate organization of an organism—a very big organism, by traditional standards. Complexity appears to rise inexorably and to pass through a threshold every once in a while. It is enough to make you wonder: Is something weird going on here?

Before that question is addressed, its basis should be clarified. The truth is that there is nothing literally inexorable about genetic evolution's, or cultural evolution's, ascent of the scale of complexity. There are species that have grown less complex through natural selection, and species whose degree of complexity hasn't changed one iota for a long time. Similarly, there are human societies that have undergone no marked structural change for long periods of time, and societies, such as the Roman Empire, that have descended from complex order into a simpler chaos.

There are at least two kinds of people who tend to dwell on examples of unchanging or descending complexity: hard-core scientific materialists who don't like the mystical overtones of the phrase "inexorably rising complexity"; and politically attuned biologists and social scientists who want to keep Darwinism from being used to justify various kinds of human sacrifice in the name of the state superorganism, or to justify the denigration of relatively simple, "primitive" human societies. Both kinds of people are sound not only in their motivation but in their conclusion: rising complexity— whether organismic or social—is not, strictly speaking, inherent in evolution.

Nonetheless, neither political sensibilities, however humane, nor philosophical leanings, however scientific, should obscure a simple and intriguing fact: *in general*, evolution has continued to create things more complex, things that shield their essential information from entropy with more elaborate means of control, and thus with more sophisticated information processing.

Further, there is reason to believe that these more complex things are more sentient, capable of more complex *feeling*. I mention this to Boulding. "There's also a rise of consciousness," I say.

"Oh, yes," he says. "Control leading into consciousness."

"Consciousness seems to accompany complex systems of control."

"Oh, yes, yes it does." In fact, he says, one fruitful way to think of evolution is as a rise in complexity, control, and consciousness. It's convenient, too: they all begin with *c*. Evolution raises the three Cs. "But why this happens," he says, "there's really no good theory on."

TWENTY-FOUR

PIERRE AND ADAM
AND KENNETH

In saying there are no good theories about the tendency of complexity to grow, Boulding is being a little hard on himself. One of the proposed theories is his, and, though it is a little on the sketchy side, it's a long way from bad. It goes roughly as follows: evolution works by filling empty niches; since the first organisms were very simple, and their niches were, by definition, already occupied, most of the empty niches were at higher levels of complexity. And thus has it been ever since: the road to novelty necessarily leads upward.

Why does Boulding give his own theory short shrift? Part of the answer may lie in his almost obsessive humility. But a probably larger, and more interesting, part of the answer has to do with the two worlds in which he lives. These worlds, and the tension between them, are reflected in the work of two men whose ideas critically influenced his intellectual development. One was a scientifically minded priest, and the other a moderately religious social scientist. Each had a theory—although one theory was more explicit than the other—about the riddle of rising complexity.

Pierre Teilhard de Chardin was born in 1881 on his family's estate in Auvergne, France. He inherited some of Voltaire's genes from his mother and memes of devout Catholicism from both parents. At eighteen, he entered the Jesuit order, and in 1911 he was ordained a priest. He was destined to be an unusual one. His childhood passion for rock collecting had developed into an enduring interest in fossils, which in turn led him to the idea of evolution and apprehension of its grandeur. Teilhard's willingness to depart from dogma in order to reconcile evolution and Christianity did not endear him to the Catholic hierarchy, and, a few years after receiving his doctorate in geology at age forty-one, he was pressured into leaving the faculty of the Catholic Institute in Paris. His Jesuit superiors suggested he accept an offer to go dig up fossils in China, where he would be safely distant from the main thoroughfares of Catholic discourse and diverted from matters of theology. There he joined the international team of scientists that unearthed the remnants of Peking Man. (Stephen Jay Gould contends that years earlier, on another expedition, Teilhard had helped perpetrate the Piltdown hoax. This accusation rests partly on an analysis of the Piltdown find, published before the hoax was exposed, in which Teilhard noted the critical absence of a piece of fossilized bone that could have determined unequivocally whether the find was authentic. Teilhard wrote, "As if on purpose, the condyle happens to be missing!" Gould, seizing on the phrase "As if on purpose," speculated that Teilhard knew more than he was saying. As we will see, Gould could hardly have found a more meaningless phrase to build his case on; the fact is that Teilhard de Chardin saw purpose everywhere he looked.)

In China, Teilhard de Chardin completed *The Phenomenon of Man*, a lyrical and moving, if often obscure, work of religious and evolutionary philosophy. The Church forbade its publication, and Teilhard abided by the ruling (although it continued to circulate among colleagues in manuscript). When the book was finally published, shortly after his death on Easter Sunday of 1955, it inspired theological controversies and something of a cult following. Teilhard associations sprang up, along with Teilhard journals and newsletters, and aspiring synthesizers began trying to fuse his thought with everything in sight: Marxism, the Montessorian method of education, and so forth. Somewhere in the Arizona desert is the Arcosanti Project, a residential community designed by Paolo Soleri and supposedly based in some sense on Teilhardian principles.

Within academia, Teilhard de Chardin's legacy has been less robust. Though his work lives on in theology courses, scientists and philosophers, by and large, do not take him seriously. Sir Peter Medawar has written that *The Phenomenon of Man* is "nonsense tricked out with a variety of tedious metaphysical conceits." Teilhard de Chardin, he added, "can be excused of dishonesty only on the grounds that before deceiving others he has taken great pains to deceive himself." But Teilhard has had his eminent scientific defenders, among them the British biologist Julian Huxley, who wrote the preface to *The Phenomenon of Man*, and the geneticist Theodosius Dobzhansky, who served as president of the American Teilhard de Chardin Association. Kenneth Boulding, too, was quite taken by Teilhard de Chardin's work when he encountered it, in the 1960s. Though not willing to defend every sentence in *The Phenomenon of Man*, he considers it "a great poem."

Teilhard's philosophy begins with the observation Boulding made over dinner: evolution tends to create beings of ever greater complexity ("complexification," Teilhard called this tendency). By evolution, he meant not merely genetic evolution but also the inorganic "evolution" that had preceded it and the cultural evolution that it had ushered in; all were manifestations of the same principle, in Teilhard's eyes. Thus, inorganic evolution created planets, like earth, and built the complex molecules that, through genetic evolution, formed cells that formed societies that congealed into organisms. Organisms then formed societies of their own, and these societies grew more intricate and cohesive themselves, in some cases under the influence not just of genetic but of cultural evolution.

Take human society, for example. Like Boulding, Teilhard de Chardin noted that the human species seems to be sustaining evolution's basic direction—that human organizations are growing in size and complexity, and that this century they have been doing so rapidly. He was especially taken by the globalization of organization. Multinational corporations sprang up profusely after World War II, and, as trade expanded, even smaller companies made enduring connections with foreign firms. There was also the postwar rise in the number of international—in theory, "supranational"—political organizations, many of them affiliated with the newly born United Nations.

Being conglomerations of organic matter, all of these organizations had to handle information—both to stay intact and to maintain vital contact with other organizations. Teilhard de Chardin, like Boulding, perceived this centrality of communication to coherence, and he placed great faith in it. As the web of organizational information crossed national borders in greater density, he believed, the peoples of the world were being drawn together. Indeed, the cultural obstacles to such convergence appeared to be dissolving, also under the power of freely flowing information; national bodies of popular culture were linked by phonograph albums and movies and, toward the end of his life, by television.

It was not too much of an exaggeration, Teilhard believed, to say that the new lines of communication formed "a generalised nervous system, emanating from certain defined centres and covering the entire surface of the globe." Mankind, he wrote, is "coming gradually to form around its earthly matrix a single, major organic unity, enclosed upon itself."

The comparison of human society to an organism is, of course, nothing new. Ever since Aristotle, people have talked about the body politic, and drawing detailed parallels between societies and organisms was a favorite pastime of some nineteenth-century sociologists (notably Herbert Spencer, who gave the enterprise a bad name by associating it with his inhumane system of ethics). Nor was Teilhard de Chardin the first person to say that the metaphor of a *global* organism had acquired merit over the years. But he was distinctive in his insistence that the metaphor had grown dramatically more apt in recent decades and was nearing a quantum leap in aptness, a leap that would have uncanny consequences.

His conviction about this coming watershed, though grounded partly in mystical revelation, grew also out of his observations of human society. And, while his extrapolation from these observations may have been lacking in temperance, the basic trajectory was not itself misguided. Even casual reflection suggests that the past century of human social evolution—including, especially, the evolution of information technology—has brought much greater validity to the superorganism metaphor.

Take, for instance, the intergenerational transmission of information. One reason that William Morton Wheeler found it so easy to think of ant colonies as organisms is that the ants, like the cells in an organism, have centralized the sending of genetic information to the next generation; all genes pass through a queen—a "winged and possibly conscious egg." Human beings, for a variety of good reasons, have not opted for that approach to genetic reproduction. But in the case of human beings, the genes do not carry all the vital information handed down from one generation to the next. The morphology of human societies depends critically on cultural information as well, and the intergenerational transmission of cultural information has been considerably centralized in modern societies.

During most of human history, the education of children came overwhelmingly from parents. In fact, it was not until after World War I that the majority of high-school-aged Americans attended high school regularly. Today about nineteen out of twenty do. Similarly, between 1940 and 1980, enrollment in American institutions of higher learning grew from 1.5 million to 12.2 million. More recently—with the popularity of day-care centers and all-day kindergarten—the trend has been to more fully institutionalize the other end of the education cycle.

Gathering twenty or thirty children in a room a few blocks from home may not seem like radical centralization, but it should be remembered that schools thousands of miles apart often use the same textbooks—and the same filmstrips, video and audio cassettes, and instructional computer programs. Now that information of all kinds is so readily replicated, the centralization of its transmission needn't imply geographic proximity.

Of course, even amid mass education, it remains true that only a fraction of culturally transmitted information travels via classroom; a biology course doesn't come near teaching all there is to learn about life. But the *extracurricular* transmission of information has also been greatly centralized, beginning around the turn of the century, with the advent of the mass-circulation magazine and the motion picture, and continuing through the present electronic bombardment. When parents complain about having lost control of their children, they are using exactly the right word; *control*—programming of the human brain with cultural information—has become less a parental responsibility and more a societal one. This ceding of power to the new class of de facto control experts—television producers, rock stars, professional athletes—may and may not have good effects, but, in its concentration of information transmission, it is decidedly organic.

Another thing organisms are known for is specialization of their constituents; hearts don't solve algebra problems, and brains don't pump blood. In human society, specialization has been growing since before recorded history, but it has been on a rampage for the past several centuries. To be sure, in the early stages of the rampage, this division of labor had little to do with information technology. It was the processing of materials, of *matter*, that the industrial revolution divided into narrower and thus more readily routinized tasks. But toward the end of the industrial revolution—and the beginning of the organizational revolution—the growing scale on which materials were being processed and distributed called for commensurate innovations in the processing of the *information* that controlled them. (This epic societal transformation has been described and analyzed by the sociologist James R. Beniger in *The Control Revolution*.)

"Bureaucracy" is the name given to the resulting division of intellectual labor. Bureaucracy originated around five millennia ago, with the governance of cities, but it reached a wholly new scale during the industrial revolution, and during this past century, fueled by the telephone and other new information technologies, its expansion has, needless to say, continued. A very rough but useful index of its size—the percentage of American workers who make their living primarily by handling information—rose from 13 percent in 1900 to 42 percent in 1960, when this "information sector" displaced the manufacturing sector as the nation's main source of employment. (There is much overlap between the information sector and the much-discussed services sector—lawyers, actuaries, and bank tellers, for example—but the two are not synonymous; auto mechanics perform a service but not mainly an informational one, while executives at Chrysler fall in the information sector but not the services sector.)

From a Teilhardian perspective, the growth of bureaucracy, and of the information sector, is viewed as nothing less than the emergence of a vast societal brain. This may sound like evidence of fuzzy thinking, but, seen against the backdrop of the history of life, the analogy acquires a certain crispness. Through genetic evolution, DNA, the original information processor, built brains and entrusted them with much of the information processing that keeps it intact. The result was a collective responsibility: a single brain processed information on behalf of the billions of copies of DNA in its organism. Then, through cultural evolution, brains built bureaucracies and entrusted them with much of the information processing on which the brains' (and thus the DNA's) survival had come to depend. For instance, the bureaucracies that keep supermarket chains running—the loose bands of vice presidents for marketing, store managers, accountants, and the rest—permit shoppers to almost mindlessly gather a diverse assortment of complex foods. Similarly, government bureaucracies relieve citizens of the need to think about garbage disposal, snow removal, and street repair. And, once again, the new level of control is collective: a single bureaucracy processes information on behalf of many brains, somewhat as a single brain processes information for many copies of DNA.

The superorganism metaphor grows more vivid, if not necessarily more valid, when labor, having been divided, is automated. For two hundred years, machines—the "mechanical phyla," in Teilhard's terms—have been acting increasingly as society's muscles, taking tasks that once entailed sweat and strain and doing them not only without fatigue but on a superhuman scale. "To an increasing extent," Teilhard wrote, "all the machines on earth, taken together, tend to form a single, vast, organised mechanism." It is sometimes said these days that the computer represents the final act of automation; having mechanized the processing of matter and energy, we are now mechanizing the processing of information. Teilhard applauded the computer's ability to "relieve our brains of tedious and exhausting work," and to "enhance the essential (and too little noted) factor of 'speed of thought.' " Had he taken more seriously the present prospect that computers may assume even some of the highest cerebral functions, he might well have beaten Ed Fredkin to the assertion that "our mission is to create artificial intelligence." And he would have meant it at least as literally as Fredkin does.

So far, this mission is a long way from accomplishment. Artificial intelligence, while giving doctors, engineers, and lawyers *some* cause for insecurity, has not nearly justified the hopes and fears its publicists generated in the early 1980s. But computers can add weight to the superorganism metaphor without doing anything very dazzling. After all the talk about automated medical diagnosis, we may not find much glamour in a machine's mere ability to store lots of information, automatically transmit and receive it, and manipulate it in elementary ways, but in just such behavior lies much of the work that human paper shufflers—in governments, supermarkets, and all sorts of large corporations—have traditionally done. Bureaucracy—the work of the *societal* brain—is gradually being automated, whether or not the most demanding work of individual brains is.

In this sort of automation lies great political, not just economic, significance. Computers can compare a list of divorced fathers whose child-support payments have lapsed with a list of males due to receive state income tax refunds and, having located the intersection, divert money from delinquent fathers to their children. Similarly, government computers now compare student loan lists with draft registration lists to ensure that America doesn't pay for the education of anyone who isn't ready to die defending it. Interconnected computers can also induce the payment of parking tickets, keep pilot's licenses out of the hands of convicted drunk drivers, and make it hard for felons to buy guns.

In all of these cases, the most striking analogy with the organism lies not in the linkage of computers—though this is indeed suggestive of neuronal networks—nor in the automation of bureaucracy, but rather in the attendant narrowing of the range of individual behavior. In organisms, after all, the behavior of constituents is closely circumscribed: human organs and cells assiduously follow neuronal and hormonal instructions; there is no such thing as a whimsical expanse of liver tissue (except, as Boulding has noted, in the case of cancer). Indeed, this rigid conformity is one of the rigorous senses in which organisms embody order: there are correlations not just among the molecular compositions of different regions but also among the behaviors of different regions, even regions separated by great distance. Thus, the organism's functioning, as well as its constitution, admits to concise summary.

Traditionally, Americans have taken pride in the poorness of the comparison between citizens and cells. They point out that in this country, unlike the Soviet Union, or Italy under the Fascists, citizens are essentially free from state monitoring. The day may come when such claims have a nostalgic air; their truth may diminish as computers, by making it harder to flout the law, erode societal entropy, whose flip side is liberty.

That is not to say that automated social order is always bad. Only a fairly dogmatic libertarian could get very worked up about transferring money from fathers to the children they've deserted. But computers can also pursue criminals less delicately and less discriminately. In addition to detecting violations of the law that are already matters of public record, they can follow the trail of private information that people leave in their hands—credit card receipts, bank statements, phone bills. And in the absence of unambiguously documented lawbreaking, machines can look for patterns of information that are merely good grounds for suspicion. New York City officials have decided to have their computers search through vehicle registration and income tax files for people who drive surprisingly pricey cars, given their reported incomes, and who thus might be fruitfully audited. Such suggestive correlations are all over the place. Maybe a particular series of phone calls—to Miami, then to New York, then to São Paulo—has a likely link with drug smuggling. Maybe large post–Super Bowl bank deposits are often made by bookmakers—and withdrawals by loan sharks. Should these clues be cause for investigation? It probably is true (or will be soon) that prosecutors, by programming computers to find shady patterns of information, could lower the number of tax dollars spent per felony conviction. Would that be justification?

And what about the similarly oppressive hand of private enterprise? Some companies have begun automatically monitoring the productivity of employees who work at computer terminals (taking airline reservations, say, or processing insurance claims) and thus leave an ongoing record of their labors. Is the growth in corporate efficiency worth the haunting sense of unceasing surveillance?

In organisms, such issues are never addressed. Genetic evolution doesn't even pay lip service to cellular autonomy unless such autonomy somehow helps the genes get safely to the next generation. The question human society will increasingly face is how far its priorities should differ from those of natural selection: how much order it is willing to sacrifice for liberty, or liberty for order. This question may sound stale, and it indeed dates back to the origin of politics, but new information technologies, by making the regulation of individual behavior cheap and easy, give it new moment. The science of surveillance and control has already outstripped George Orwell's imagination; we have the machinery to turn the United States into a state organism so taut as to make the Oceania of *1984* look like an earthworm by comparison. And even in Oceania, O'Brien, the totalitarian goon, seems to have the equation basically right when he says to

Winston Smith, whose identity he is methodically crushing, "Can you not understand, Winston, that the individual is only a cell? The weariness of the cell is the vigor of the organism."

Teilhard de Chardin's critics have charged that his vision of a superorganic society implies just such individual sacrifice, but he insisted that this trade-off is illusory. "Individualization" and "aggregation," he said, are not only compatible but in some ways inseparable; we realize our identities most fully when devoted to a greater whole. As for the totalitarian societies that seemed to belie this rule—such as the Soviet Union, where he found "the anthill instead of brotherhood"—the problem lay in "perversion" of the laws of evolution; a more enlightened social design, more cybernetically sound, was the solution: "When an energy runs amok, the engineer, far from questioning the power itself, simply works out his calculations afresh to see how it can be brought better under control." Alas, Teilhard did not go into much detail about these calculations.

All told, this idea of human society as an organism is very useful. First, it suggests a thumbnail summary of the significance of the information age: during the past hundred or so years—beginning with the coming of the telephone and continuing through many other breakthroughs in information technology, including the microcomputer—the superorganism metaphor has gained more validity than during any previous hundred years in history. Second, this thumbnail summary frames one of the half-dozen or so great political questions of the coming few decades: How organic do we want society to become?

Actually, there is one other looming political question that also lies within the superorganism metaphor. As we've seen, Teilhard was not looking just at the United States, or France, or Russia, as an organism. He would agree, to be sure, with anyone who characterized them as such, but he was much more taken by the idea of a single social organism enveloping the entire globe. Here the comparison is not so much between individuals and cells as between nations and cells; the nations steadily surrender increments of their autonomy to one another and thus to the larger, collective order.

In the wake of World War II, as in the wake of World War I, the political unification of the planet was the subject of much earnest discussion. Among its advocates, opinion varied greatly on the question of its ideal degree. Is a world state, with its dangerous monopoly on power, really necessary, or would a global federation of semi-autonomous states suffice to keep the peace? These questions, recast in organic terms, amount to questions about the structure of the earth's nervous system. In a world state that had resulted from conquest, for example, overall control would probably be, as in human beings, confined mainly to a single area. But if unification came through a slow drift toward federalism—through, say, the Order of World Citizenship envisioned by the young Kenneth Boulding, or even through a strengthened United Nations—the world's network of political communication would look more like the nervous system of a jellyfish; control would be widely distributed and nowhere very rigid.

Of course, such images now seem a little on the fantastic side. After four decades of nuclear standoff and Cold War, world conquest and a substantial symmetrical surrendering of national autonomy appear—at first glance, anyway—to be in a dead heat for least likely scenario. The point for now is simply that global political unification, regardless of its prospects, is an idea tied to the image of the superorganism; it is raised by extrapolating from the increasingly organic structure of society, and the questions it poses can be illustrated in organic terms.

All things considered, then, the superorganism metaphor is worth contemplating. It captures some important trends and raises some good questions about the future of nations and of the world. Indeed, the importance of at least some of these questions—questions of societal efficiency and personal liberty, for instance—has grown since Teilhard's death. If there were an award for most prescient use of metaphor, the award for the information age would have to go post-humously to Pierre Teilhard de Chardin for "society as organism."

Teilhard de Chardin, though, would have to decline the award posthumously on grounds that he had not meant to be taken metaphorically. He believed that the human species was in some fairly literal sense an organism, and that this sense would become dramatically more literal in the near future. To grasp his meaning—or to come as close to that as possible without having a mystical experience—you must understand that his conception of evolution, like Boulding's, involved a kind of dualism. He believed that what appeared outwardly as a rise in physical complexity was matched, "within" each organism, by a corresponding growth in consciousness; in particular, the complexity of an organism's information-processing system seemed to have an especially close connection to consciousness.

In talking about the "consciousness" of, say, a bacterium, Teilhard was not referring to self-awareness; he didn't mean that bacteria sit around and wonder if there is an afterlife. He just meant that they are sentient—that (to use the refreshingly lucid terminology of the philosopher Thomas Nagel) it is *like something* to be a bacterium. It isn't like *much*, probably, because bacteria aren't very complex, but it is like something, Teilhard would say. Worms, he would say, have a bit more consciousness, and dogs, relatively speaking, have a lot. Humans, being the most physically complex animals—and, moreover, having the most complex nervous systems—lead the league. Humans, indeed, have not only consciousness but self-awareness. Not only is it like something to be them; they *know* that it is like something to be them, and they discuss the philosophical implications of this fact.

If consciousness indeed grows along with physical complexity—and especially with the complexity of information processing—what happens when individual humans are woven with information technology into a single, unfathomably complex organic fabric that envelops the world? Well, said Teilhard, their shared body of thought can then be said to constitute a "noosphere"—a kind of global brain, the collective repository of all nongenetically transmitted information. (Boulding, virtually alone among social scientists, uses this term; he refers to cultural evolution as noogenetic evolution.) To the extent that people the world around are plugged into the noosphere, especially via such instantaneous media as radio and television, there exists "a sort of 'etherised' universal consciousness," Teilhard wrote.

By "universal consciousness," he apparently meant several different things, depending on context. At times he seemed to mean merely what Marshall McLuhan (who had read Teilhard de Chardin) would gain fame for saying years later about the "global village." In "The Planetisation of Mankind," an essay written in 1945, Teilhard observed that, since the beginning of the war, "economically and psychically the entire mass of Mankind, under the inexorable pressure of events and owing to the prodigious growth and speeding up of the means of communication, has found itself seized in the mould of a communal existence—large sections tightly encased in countless international organisations . . . and the whole anxiously involved in the same passionate upheavals, the same problems, the same daily news." Here he does not appear to be saying anything of metaphysical depth; it doesn't sound as if he means that consciousness has been literally unified—that it is *like something*, like *one thing*, to be the human species. Rather, he seems merely to be talking about long-distance interpersonal awareness: Londoners are aware of the plight of Parisians (and often share their plight), and vice versa. And at times he went just a bit further, suggesting that such mutual awareness would bring mutual "sympathy."

If this were all Teilhard meant by universal consciousness, his mystic vision would be, if not exactly convincing, at least conceivable. Maybe middle-class Americans will come to empathize with the refugees and earthquake victims whose images they see on the evening news. Maybe millions of copies of the album *We Are the World* will forge lasting international bonds. Maybe the Russians will be able to muster some sympathy for Americans once they've seen twenty or thirty episodes of *Gilligan's Island*.

But Teilhard had something more than this in mind. At times he spoke more expansively about the meaning of universal consciousness, and made it clear that he wasn't referring to the sort of sympathy conveyed by Hallmark cards. He was really talking about *love*, the kind of altruistic diffusion of identity traditionally reserved for kin. Indeed, he was talking about even more than that. Love was not, in Teilhard's scheme of things, a mere by-product of human evolution, an emotion programmed into the brain to ensure the survival of the DNA. Rather, it was a manifestation, the most important manifestation, of the "spiritual energy" that had been growing in the "within" of ever-more-complex matter since before the creation of life, and that continued to grow not just through genetic evolution but through cultural, or noogenetic, evolution and the "complexification" of human society.

As this tide of spiritual energy reached some critical level, and brotherly love achieved global scope, the phrase "universal consciousness," Teilhard believed, would take on the strangest of its several meanings. Eventually—and this is where he parts company with just about everyone, including Boulding—human consciousness would become *literally* collective, subjectively collective; spheres of individual consciousness would fuse into a single, worldwide "hyperpersonal consciousness." Humankind would then reach "Point Omega," the ultimate realization of evolutionary potential.

Depictions of Teilhard de Chardin's philosophy are necessarily sketchy. Like most mystics, he had trouble articulating his vision. It is not clear exactly what Point Omega was, or what life would be like after it had dawned on the planet. Complicating the notion of "hyperpersonal consciousness," for example, is the fact that Teilhard believed—in keeping with his insistence on the compatability of "individualization" and "aggregation"—that individual realms of consciousness could be preserved even as they dissolved into a larger consciousness. He wrote: "For men upon earth, all the earth, to learn to love one another, it is not enough that they should know themselves to be members of one and the same *thing*; in 'planetising' themselves they must acquire the consciousness, without losing themselves, of becoming one and the same *person*."

In the end, confronted with Teilhard's cherished paradoxes and sometimes opaque imagery, the most we can say with complete confidence is that he saw the culmination of human evolution as being something out of the ordinary: "Not disintegration and death, but a new break-through and a re-birth, this time outside Time and Space, through the very excess of unification and co-reflexion." (His writing, rife with quirky terminology, loses something in translation, according to his bilingual devotees.)

Teilhard's habitual obscurity clouds a question of particular interest: What did he see as the prime mover in "complexification"? Is the growth of organic complexity driven by the logic of physical evolution, while consciousness expands alongside it in accordance with some metaphysical law equating the two? Or does spiritual energy, with which he identified consciousness and love, grow independently, impelled by some mystical momentum, and weave molecules into the structures that house it?

In one sense, this question doesn't matter. Teilhard de Chardin definitely saw evolution as teleological. God's will was behind the whole thing, one way or another. But in another sense, the question matters greatly. At issue is whether Teilhard could be convicted of first- or second-degree teleology. The idea that God is behind evolution is technically permissible, if not especially popular, in scientific circles, so long as God is supposed to work aboveboard and without periodic intervention—so long as He simply wound up the great universal clock, as the deists of the eighteenth century believed, and let it run. Thus it was permissible, technically speaking, for Ed Fredkin to step back from the scientific domain and suggest, in a metaphysical vein, that some sort of intelligence set the universal computer in motion to solve some problem. But scientific respectability will be hard for Teilhard to come by if he believed that God works in a weirder way; that the "within" of organic matter—its subjective interior, home of consciousness and the experience of love—draws people (and cells, and protein molecules) together like magnets.

For the most part, Teilhard's explanation was of this more suspect sort. Though he occasionally paid his respects to the physical mechanisms of evolution, according some measure of credence to natural selection, he did not find it sufficiently powerful to account fully for the evolutionary trajectory toward complexity and integration. The more basic impetus, he suggested, lies within, in the rising tide of "spiritual energy." "According to current thought, an animal develops its carnivorous instincts *because* its molars become cutting and its claws sharp," he wrote. (Actually, this is not a perfectly accurate rendering of thinking on the matter—then or now.) "Should we not turn the proposition around? In other words if the tiger elongates its fangs and sharpens its claws is it not rather because, following its line of descent, it receives, develops and hands on the 'soul of a carnivore'?" He wrote elsewhere, "The impetus of the world, glimpsed in the great drive of consciousness, can only have its ultimate source in some *inner* principle, which alone could explain its irreversible advance towards higher psychisms."

As evolution progresses through its "noogenetic" phase, this inner principle grows in importance; the "within" of things, according to Teilhard, gradually comes to dominate the "without." So, during those last few taxing miles to Point Omega, spiritual energy is a more vital fuel than ever. "Nothing seems finally capable of guiding us into the natural sphere of our inter-human affinities except the emergence of a powerful field of internal attraction, in which we shall find ourselves caught *from within.*"

Teilhard de Chardin had, as Philip Hefner, a theologian and the author of *The Promise of Teilhard*, has put it, a "near obsession for unity and synthesis." Now we see how lavishly he indulged his obsession. To begin with, he could, like Ed Fredkin, E. O. Wilson, and Kenneth Boulding, enjoy the Ockham epiphany: the apprehension of a single principle that unifies diverse phenomena. In his case the principle was not quite as neat as people like Wilson and Fredkin like their principles; the logic he saw behind evolution was a tendency of spiritual energy—a messy concept, at best—to grow, and to carry physical complexity along with it, perhaps abetted in some vague way by natural selection. But there is a kind of amorphous unity in this explanation, and thus, for Teilhard, a kind of satisfaction.

More to the point, though, Teilhard could also enjoy the epiphany of *teleological* explanation; he had found a unified explanation not just of how but of *why* evolution works: it is a divine means to an end, a premapped route to spiritual integration. If the epiphany of scientific explanation is like reaching the climax of a mystery novel, when a single cause is found for various mayhem, then this epiphany of teleological explanation is more like reading an allegorical novel; it is like the unscrambling of symbols that permits the reader to understand not just how the plot works, but why it works that way—the book's purpose, the author's intent, the whole point of the exercise.

As if these two epiphanies weren't enough, Teilhard de Chardin's philosophy offers an aesthetic bonus, a glimpse of a third kind of unity. For the author's intent—the unifying teleological explanation of life, evolution, and human experience—turned out to be, yes, unity: physical, psychic, and spiritual unification of the planet. This, Teilhard's third epiphany, has some of the earmarks of the classic mystical experience: a sense of the oneness of everything, a sense that this oneness is the answer to the Big Question, and the stubborn resistance of the answer (and even the question) to clear articulation.

It is difficult for people not privy to mystical experiences to imagine their power, but my guess is that most of us get just a whiff of them now and then. There is certainly something aesthetically moving about watching the fragile cohesion of a soccer team solidify as the players form fluid and subtly organic patterns that sweep the ball toward its destination. Those not athletically inclined may prefer to watch a flock of birds cross the sky at twilight; upon changing course it will fan out and then quickly coalesce, tenuously preserving and suddenly consolidating its unity. It is easy to imagine at such moments that the very purpose of those birds' lives has been to get them to this point, when they can lose themselves in the beauty that together they constitute.

But of course, so far as science can tell us, that is not the reason the birds were born. So too with evolution. So far as we know, our purpose is not to form a global organism—and we needn't resort to such teleology in explaining how humans came to be, or why organic complexity keeps rising. We can account for these things—in rough outline, at least—with a conventional, concrete explanation, grounded in the laws of evolution.

Some of the central themes in any such explanation can be found in the work of Adam Smith, the Scottish economist who attended Oxford in the 1740s and whose ideas Boulding encountered there nearly two centuries later. Smith never explicitly addressed the question of rising organic complexity. And it would be unreasonable to expect this of him; he died in 1790, nineteen years before Charles Darwin was born. But Boulding nonetheless considers Smith "the first evolutionary economist." (In homage, he owns a T-shirt featuring Smith's visage.) And, indeed, Smith's writing about economic complexity assumes great relevance to biological complexity if you look at it with general system theory in mind and ask what economic systems, like corporations, have in common with biological systems, like multicelled organisms.

One answer is that both must compete for survival, and in both cases the bottom line is efficiency. Less efficient rabbits starve or get eaten, and less efficient corporations file for protection under Chapter 11. So, one logical thing to expect of evolution—whether it is the genetic evolution that created organisms and societies or the cultural evolution that shapes social institutions—is that it move organizations toward greater efficiency.

In *An Inquiry into the Nature and Causes of the Wealth of Nations*, which has been the bible for the American economic system since its publication in 1776, Smith isolated one of the main ingredients of efficiency. He wrote, in the first sentence of the first chapter, "The greatest improvement in the productive powers of labour, and the greater part of the skill, dexterity, and judgment with which it is any where directed, or applied, seem to have been the effects of the division of labor." Smith's famous illustration is the pin factory. "One man draws out the wire, another straights it, a third cuts it, a fourth points it, a fifth grinds it at the top for receiving the head; to make the head requires two or three distinct operations; to put it on, is a peculiar business, to whiten the pins is another; it is even a trade by itself to put them into paper." Even a small pin factory, with only ten men, can produce 48,000 pins in a day, Smith reported, whereas those same men, working independently, could make no more than two hundred. "In every other art and manufacture," he wrote, "the effects of the division of labour are similar to what they are in this very trifling one." (Kenneth Boulding, in his undergraduate notes on Adam Smith, elliptically summarized the gains in productivity due to division of labor in modern societies: "Common coat the result of combination of vast number of workmen—Humble worker better off than African King!")

Smith did not, presumably, mean by "every other art and manufacture" to include endeavors so far removed from pin making as, say, converting a roast beef sandwich into a fine pulp and extracting energy from it. Nonetheless, he probably realized that the logic of the pin factory applies equally to subsocial levels of organization; in the name of productivity, stomach, intestinal, and other cells divide the labor of digestion, just as straighters, whiters, and other people divide the labor of making pins.

The moral of the story at the two levels of organization is the same: members of a team can do more with less if they assume separate functions. Further, division of labor is a powerful argument for forming teams in the first place, and for expanding teams once they're formed; the more members, the more finely labor can be divided. There are other reasons, too, why larger systems are sometimes more efficient than smaller ones, and economists list these reasons under the heading "economies of scale." The point here is not to enumerate them but simply to note their existence; efficiency can often be improved by expanding organizations, as well as by dividing their functions more minutely. So, wherever efficiency is a central rule of design—as it is in both cultural and genetic evolution— systems of great size and complexity should be forthcoming.

We are playing fast and loose here, indulging in a fairly facile extension of principles from one level of organization to another. The truth is that there are many relevant differences between levels. Kin selection, for example, almost certainly played a more prominent role in the rise of organismic than of social complexity. But such differences are probably not sufficient to undermine the basic argument: it isn't a spooky coincidence that cultural and genetic evolution move in the same direction, but rather a dual manifestation of a single set of principles.

One of the more conspicuous differences between the two evolutions, though, does merit mention. Organisms are endowed with division of labor not by intentional innovation but by serendipitous genetic mutation; natural selection has discreetly removed from the stage the many organisms whose mutations were less serendipitous. Some of that sort of culling has gone on with human economic organizations, too, but the rise in social complexity has more to do with the fact that humans can conceive innovations expressly designed to increase efficiency and thus keep their organizations from getting culled. Moreover, other humans can see what works and copy it; they can read books about how the Japanese do things, and exchange helpful hints at conventions and during coffee breaks. In other words, culturally based division of labor, unlike genetically based division of labor, can spread intragenerationally, through emulation. Such horizontal transmission is one of the primary advantages that cultural evolution has over genetic evolution, and one of the reasons that the last six thousand years have seen a growth of complexity in human societies that, in the organisms of a species, would take orders of magnitude longer.

This is not to say that the division of human labor has always resulted from conscious understanding of its efficiency. On the contrary, it often emerges without planning, as a by-product of trade; potatoes get grown in Idaho, and apples in Oregon, without anyone giving the overall scheme much thought. This spontaneous specialization is what permits the "invisible hand" of capitalism, as Smith called it, to translate the narrow-minded pursuit of individual gain into a broader order: the efficient allocation of society's resources. Through the invisible hand, all traders—that is, all participants in a modern economy—exploit the principle of division of labor without necessarily being aware of it.

This would hardly surprise Robert Trivers. According to his theory of reciprocal altruism, some of our very distant, very hairy ancestors exchanged goods and services, like food and grooming, before they were acutely aware of *anything*. (Richard Dawkins defines money as "a formal token of delayed reciprocal altruism.") If Trivers is right, and the impulses underlying barter are rooted in our genes, it means that genetic evolution built division of labor not only into our bodies, but also, more grossly, into the structure of our societies. Cultural evolution, impelled by the pursuit of individual self-interest, then carried the social division of labor to greater definition.

All told, both kinds of evolution, genetic and cultural, have taken self-interest, whether pursued blindly by genes or more sentiently by people, and translated it into a larger order—or, more precisely, a larger complexity. Both evolutions thus qualify for the term *invisible hand*. You might say that one invisible hand has created another.

Division of labor, the various economies of scale, and whatever other factors underlie growth in the degree and scope of social complexity amount, at best, to an explanation of what Boulding has called the demand side of the complexity equation; they help explain why organic systems often evince an *impetus* toward greater size and complexity. The explanation of the supply side offered by Boulding in *The Organizational Revolution* remains valid: that impetus can express itself only with suitable equipment; the growth and "complexification" of all sorts of organizations—and their intertwining—have come courtesy of appropriate technologies for processing and transmitting information.

Indeed, so intimately related are integration and information that, viewed from a distance, they leave it unclear which is following which. If you had set up a camera on the moon a hundred years ago, and loaded it with some sort of special film capable of capturing information in transit, and photographed the earth once a month for the next century, the result would be a truly sublime time-lapse motion picture. Telegraph and then telephone lines would creep across continents like ivy inching along an earthen bank, and strands of information would leap up to satellites, drop down across the seas, and thicken like grape vines during spring. If you watched this movie from a sober and scientific, Smithian point of view, it would look as if the planetary nervous system were being assembled to accommodate an increasingly expansive division of labor. But from a mystical and metaphysical, Teilhardian point of view, it would look as if the planetary nervous system were forming of its own volition and sucking humans into its web in accordance with some divine plan.

Somewhere between these two views—or, perhaps, straddling the two of them—is Kenneth Boulding. He is compelled by his analytical bent and by vocational protocol to seek sober explanations of complexity's rise. But, by virtue of his religious convictions and his mystical sensibilities, he is not above hinting that something weird is going on here.

There is a second sense, too, in which Boulding's thought encompasses that of Smith and Teilhard de Chardin. He has worked the spirit of both men into his grand design, his theory about the three great systems that hold society together. Boulding's "exchange system"—the economy—was Smith's area of expertise, and in it, Smith showed, lies social cement, an interdependence among people and among groups grounded solidly in self-interest. Boulding's "integrative system," with its churches and charities and family reunions, is the medium of love—of the "sympathy" that Teilhard de Chardin was counting on to pull humanity toward Point Omega. (Coincidentally, Adam Smith, too, was very taken by the power of human sympathy. He began his academic career as a moral philosopher, and in *Theory of Moral Sentiments*, his first book, he defined sympathy in these terms: "Whatever is the passion which arises from any object in the person principally concerned, *an analogous emotion* springs up at the thought of his situation, in the breast of every attentive spectator." Smith believed the sympathetic impulse to be innate; in 1964, when the theory of kin selection was articulated, this belief found its most rigorous theoretical support yet.)

The importance Boulding attaches to the integrative system is best seen in relation to the third part of his trinity—the "threat system." The threat system has a long history. Shortly after agriculture and shepherding were discovered, societal surpluses began to accumulate, permitting some workers to spend their time doing something other than finding food; the seeds of the leisure class, of the exchange system, and of the first cities had been planted. But according to Boulding, it was through coercion, not through exchange, that the inchoate ruling class first convinced everyone else to part with surplus food. So the threat system, he says, was, very near the beginning, integral to the "complexification"—the incipient urbanization—of society. And it continued to play a big role for some time thereafter; the geographic scope of complex societies was expanded not by negotiation but by the wanton conquest of less complex societies. The United States, for example, carried political organization in North America from the tribe to the nation-state largely through violence and intimidation.

Boulding is happy to report that things have changed. Societies increasingly are held together not by threat but by exchange. Nations behave civilly toward one another less and less from fear of conquest and more and more out of a need for goods and services (or loans or loan payments). The shifting emphasis from threat to exchange is part of a long trend Boulding calls "gentling." Whereas Scandinavians once raped and pillaged, now they fish and farm. Whereas much manual labor was once performed by slaves, little is today. Whereas dueling once settled disputes, lawyers now do. Civilization, in other words, has brought a certain degree of civilization. The threat system has surrendered some of its responsibility for international and domestic order to the exchange system.

Still, it has not surrendered all of that responsibility. The threat system still helps keep some of the peace, both within and among nations, and Boulding worries about the difficulty of weaning ourselves from it. He doubts that the exchange system, by itself, can carry the extra burden of domestic and international order that would be left if all threats were withdrawn; a society glued together only by money and cold calculation may lack "that sense of legitimacy and community necessary to sustain it." The solution, Boulding believes, is to be found in the integrative system.

But this raises problems—not just because he has yet to come up with a satisfactory definition of the integrative system, but because the integrative system is a two-edged sword. It is centered around churches, lodges, and families, and the like, but it also involves all sorts of "symbols of legitimacy"; and these symbols include not just crosses, the Elks Club emblem, and wedding rings, but also symbols of allegiance to institutions that are themselves not primarily part of the integrative system. Thus, loyalty to nations, and to armies, though partly rooted in exchange or threat, is also an integrative phenomenon. The part of us that stirs upon seeing the flag is closer to love then to fear or cold calculation.

So integrative symbols are tricky things, and dangerous. Just as they can direct allegiance toward the core of an organization, they can direct hatred and violence outward. Witness the Crusades, the Inquisition, and World War II.

Is there any prospect for employing integrative symbols more benignly? Could impulses of affiliation somehow supplant threat and supplement exchange as ways of holding the nations together? Can we cast the integrative net worldwide? Such are the questions Kenneth Boulding is concerned with these days. He is in some ways encouraged by the general expansion of loyalty over the centuries; whereas emotional affiliation was once confined mainly to kin, it has grown to embrace the tribe and even the nation-state. So it could, conceivably, take the next logical step, to the global level, if it could somehow exploit modern technology—extend its reach via the thickening web of satellite beams and optic fibers, TV sets and computer bulletin boards, video conferences and space bridges. The essential challenge was foreseen by Teilhard de Chardin in 1947: "Humanity . . . is building its composite brain beneath our eyes. May it not be that tomorrow, through the logical and biological deepening of the movement drawing it together, it will find its *heart*, without which the ultimate wholeness of its powers of unification can never be fully achieved?"

The theory of kin selection suggests caution. It may be difficult to take truly brotherly love—our deepest and purest impulses of altruism, generally reserved for our nearest relatives—and apply it broadly. And, presumably the broader the application, the trickier a

maneuver this is. "There are unfortunately limits to love," Boulding
wrote in *Ecodynamics*, "and the question of how to expand these limits
is very important." He places some hope in education of the young;
one approach that he has proposed in delicately couched language is
to instill a sense of "generalized benevolence" through "the development of symbolic systems like 'the love of God.' "

TWENTY-FIVE

THE ULTIMATE SYNTHESIS

After dinner, Boulding and I adjourn upstairs, to the Penington House parlor, which is a little reminiscent of the Munsters' living room. There are no cobwebs, but the ceiling is at least twelve feet high, and most of the furnishings are a good half-century old. The Oriental rug is threadbare, the piano is beginning to slouch, and the grandfather clock, which reads 8:01, is genuine. The bookcase houses an edition of the *Encyclopaedia Britannica* that was published in 1910, the year Kenneth Boulding was born. His book of religious poetry, *The Naylor Sonnets*, lies on an end table. He and I are sitting on a sofa against the wall.

This is probably my last chance to talk with Boulding in person. He is leaving New York in a few weeks to spend the summer in Boulder. He has already spent much time answering my questions, in addition to letting me observe him in various habitats, and, notwithstanding his friendliness, generosity, and nearly infinite grace, I believe he has seen about as much of me as he wants to see for a while. So I skip the preliminaries and proceed to the big questions.

Why does he get such a thrill out of apprehending any principle that applies at different levels of organization? Without thinking long he says, "Well, it's fun." He pauses. "And it, it, uh, well"—he sounds momentarily unsure about the wisdom of continuing, but he continues—"my religious life may have something to do with this, you see—just the idea that the world ought to make sense, I mean, that it has a unity to it, and is not just a set of bits and pieces. There are these underlying unities in the midst of diversity."

Slow, arrhythmic steps are coming from the hall. Strained breathing grows in volume. Mary appears. She makes her way over to a table a few feet from us, leans her cane against it, eases herself into a chair, and watches.

Boulding continues: "The diversities are very important as well, you see. I'm a great fan of variety." This is one of the odder things about Boulding's mind: his love of theoretical unity is rivaled only by his love of empirical nonconformity. Most scholars seem to tend in one direction or the other. They spend their time either glossing over details for the sake of bold generalizations, or dwelling on the details that impede confident generalization. Boulding spends half his time erecting bold generalizations and the other half tearing them down. No sooner does he describe the threat, exchange, and integrative systems than he is stressing the hopelessness of disentangling the three. Why, I ask, must he so insistently temper his intellectual daring with humility? "Well, after all, it's a very large universe, and we're really very small," he replies. "There's a lot to be said for reasonable modesty." Besides, "One always has to keep a certain sense of mystery, a sense of the unknown, or even the unknowable. It would be very presumptuous of us to think that with our biological equipment we can know everything. The ant doesn't know much about us, certainly." Mary gets a good laugh out of this. He continues: "And there's no reason to suppose that what we can know is all there is to know. There's also, I think, a very important place for the religious experience. A sense of mystery is found in it, you see. In a sense, worship is almost the adoration of the unknowable."

Mary has decided to make a contribution to the conversation. "Arthur Clarke sounds a little bit like you," she says.

"Hmmm?" asks Boulding.

"Arthur Clarke sounds a little like you."

"Oh, yes," he says. "Yes, yes, yes."

Mary won't be dismissed so easily. "Have you ever read him?" she asks.

"Some," Boulding mumbles. "A little bit." He confesses: "I don't really know enough about him."

"He wrote *2001* and some other things," Mary says.

"Yes, he's a great science fiction writer. But—" He looks at me, visibly eager to avoid a discussion with Mary about the parallels between his work and Arthur Clarke's, and picks up roughly where he left off. "The, uh, you see, I've always felt a real community with the mystical tradition." He names a few highly esteemed mystics: George Fox, who founded the Society of Friends more than three hundred years ago; Dag Hammarskjöld; Thomas Merton. Mary says she has done some reading about Thomas Merton. Boulding adds, "And Teilhard de Chardin, of course, had a great influence on my thinking, really. *The Phenomenon of Man*, especially."

The doorbell rings. Mary gets up and goes to answer it.

Teilhard de Chardin sounds like a good way to get Boulding launched on some cosmic speculation, which is exactly what this room has put me in the mood for. I say, "Yeah, this sort of inexorable rise of consciousness certainly seems like reason to think that something uncanny is going on here."

He doesn't take the bait. "Yes," he says simply.

The time has come, I decide, to talk with Kenneth Boulding about the weirdness of consciousness. The weirdness of consciousness first occurred to me a couple of years ago, probably during one of my periodic attempts to find evidence that life is not devoid of meaning. On numerous occasions since then, I have tried to explain this weirdness to other people. Most have looked at me oddly and either nodded vaguely or confessed that they couldn't figure out what I was talking about. Those who have shown signs of understanding what I was trying to say have, after reflecting on it, said things like "So?"

These reactions are all the more troubling because the weirdness of consciousness strikes me as so clear and important. The logic leading to an appreciation of it seems as cut and dried as the Pythagorean theorem, and much more consequential. So if I am wrong about the weirdness of consciousness, I am probably suffering from serious conceptual distortions, the kind that typically precede talking to household appliances and taking advice from dogs. Kenneth Boulding's opinion thus assumes great significance. It is within his power to reassure me that I have not yet lost my bearings. He is not the best-qualified judge of sobriety, perhaps, inasmuch as he was declared ineligible for the draft on psychiatric grounds forty-five years ago, shortly after admitting that, yes, in *some* sense, he sometimes hears the voice of God. Still, I am in no position to be choosy. Everyone else has failed me.

If you are trying to grasp the weirdness of consciousness, it helps, ironically, to be a hard-core determinist, a determinist of the old-fashioned, Ed Fredkin/Albert Einstein kind—to believe that everything that happens, including all human behavior, is inevitable; that the future could in principle be precisely predicted, given the present state of the universe and the laws that govern it. If you are this sort of determinist, then you believe that free will is a myth; you believe, in direct disagreement with Pierre Teilhard de Chardin, that consciousness—the subjective side of reality, the world of feelings and thoughts—has no causal role in human behavior; and you therefore believe it is possible, in principle, to give a complete explanation of a person's behavior by referring only to physical things. Instead of saying Jack fled out of fear, you would say his fleeing was caused by the interaction of epinephrine, neural impulses, and various other tangible forms of information whose flow corresponds to fear. By "corresponds," you would mean that these flows of information *cause* the sensation of fear as a side effect, at the same time that they are causing the behavior of fleeing—but that the sensation of fear does not, in turn, cause anything. In this view—the view of your garden-variety determinist and reductionist, the kind of person who could loosely be called a scientific materialist—sensations are like the shadows in a shadow play; the subjective world is affected by, but does not affect, the physical world.

The same could be said of more cerebral sensations, such as the feeling of "figuring out" an answer in a crossword puzzle. This feeling is not what causes a person to write down the answer; the same neuronal processing of information that leads to the feeling is what causes the person to write down the answer. The feeling is just thrown in at no extra charge, and with no further effect.

In short, according to old-fashioned determinists, consciousness—the realm of sensations, of subjective experience—doesn't *do* anything; it is a mere epiphenomenon. Granted, it is common for biologists to say such things as "The pleasure that sex brings is evolution's way of getting us to have sex." But if these biologists are truly scientific materialists, and truly determinists, they don't mean what they're saying; they mean that the flow of physical information leading to pleasure—a flow that gets us to seek sex just as surely as the flow of information in a toilet gets it to "seek" fullness—is evolution's way of getting us to have sex. If, through some sort of experimental metaphysical legislation, consciousness could be neutralized—if our bodies were constituted just as they are now, only it wasn't *like anything* to be a human being—we would still function normally, according to this view. There would be marriages, with their vows of devotion and their behaviors of mutual obligation, but no sensations of love; wars, with their violence, but no feelings of hatred.

All of this leads to the big question: Why *does* it feel like something to be a human being?

The depth of the question is best understood in the context of natural selection—not the natural selection that created human beings, necessarily, but natural selection in the abstract. Consider a generic, lifeless planet in another corner of the universe. Suppose that, for some reason, some of its molecules start producing copies of themselves, and that these copies do the same, as do *their* copies, and so on, ad infinitum. Copying errors are occasionally made, and, by definition, those errors conducive to the survival and replication of the resulting copies are preserved, whereas errors that are not so conducive are not. It so happens that a string of copying errors, guided by this selective pressure, leads to the encasement of some replicating molecules in little cellular houses. In similar fashion—through the selective preservation of mutations—additional layers of protection are added; these houses are integrated into huge housing complexes—mobile housing complexes, no less, complexes that lumber around the surface of the planet. And, necessarily, these complexes handle meaningful information; they absorb molecules—or photons or sound waves—that represent states of the environment and that induce behaviors appropriate to those states. Indeed, this information is sometimes exchanged; one housing complex sends representations to another complex, and these symbols, upon their arrival, induce elaborate chains of internal activity that culminate in appropriate behaviors.

Now, is there any reason to believe that it is *like anything* to be one of these cellular complexes? Of course not. So far as we can see, these are mere automatons, mere robots; there is no reason to expect them to be anything else. It is pretty difficult to imagine a mutation that would endow them with the capacity to experience sensation, and, moreover, it isn't clear how such a mutation could help them; everything sensations might seem capable of accomplishing can also be accomplished through the movement of physical information.

This description of evolution could—surprise!—be applied to planet Earth. Indeed, most evolutionary biologists would endorse it as a generally accurate description of how we came to be. Such endorsement amounts to implicit agreement that there is no obvious reason for any of us to be conscious: the phenomenon of subjective experience is evolutionarily superfluous.

All of this would be less noteworthy if consciousness were just another feature, like five-toed feet or whitewall tires. But consciousness—sentience—is precisely what gives life at least a modicum of meaning, and morality a basis. The reason life is worth living is that it has the potential to bring pleasurable sensations—love, joy, etc. The reason it is wrong to kill people is because death deprives them of future happiness they might otherwise have experienced, and because it causes their friends and relatives to experience pain. If there were no such things as pain and happiness—if it weren't like anything to be alive—what would be wrong with knocking off a few humans on a Saturday night?

So this is what I find so weird about consciousness: the very thing that gives life a kind of meaning is the thing that the theory of natural selection doesn't quite explain. And this conclusion—which sounds suspiciously like something that would come out of Jerry Falwell's public relations office—is in fact the product of good, old-fashioned godless determinism. Ironic, no? And shocking, too, if you, like me, have spent much of your life assuming that the theory of evolution pretty much settles every basic mystery about life, with the exception of the origin of self-replicating molecules.

This is not to say that the theory of natural selection is flawed as an explanation of how humans came to be; my devotion to the theory is not diminished by the weirdness of consciousness (though my awareness of the bounds of its power is sharpened). It is just to say something that I guess most people take for granted anyway: life in this universe is a strange thing.

Now for some comic relief: I try to convey concisely the weirdness of consciousness to Boulding. Summoning the full body of my rhetorical skills, I begin: "I mean, there's no, I believe there's no reason . . ." I start again: "If you were a strict scientific materialist, uh, I mean, there's no reason to expect consciousness to take place, there's no need for . . ." I start again: "In a strictly scientific materialist framework, there's no need to invoke consciousness as an explanatory, I mean, there's no . . ." I start again: "Evolution can explain everything physical about us, but it doesn't seem to me to explain why it feels like something to be alive. We could be robots—function just as intricately, and our behavior could be just as intricate theoretically— and it needn't feel like anything to be us."

In an act of pure Christian charity, Boulding acts as if I'm making myself clear. "No," he agrees, it needn't. "Well," he continues, "just the fact that we have this access, you see, to the mind, which we don't understand at all. Well, there's no model of consciousness, is there? We don't have the slightest idea how we would build a conscious computer. . . . It's extraordinarily puzzling."

Having endured my tangent, Boulding now goes off on one of his own. "It's certainly clear to me that evolution primarily is a process of—well, after all . . . matter and energy are mainly significant as transmitters and coders of information. I've always pointed out what happens in a conversation like this, where something starts in the nervous system—certainly structures of some sort. These are translated into electrical impulses to the vocal chords, and these are translated into physical movements of the vocal chords, and these are translated into air waves, and then these hit your ear, and they're translated into nervous impulses that go to your brain." He points to his head and then to mine: "And some structure here turns into some structure there, you see, with all these innumerable intermediaries." The moral of the story: "The structure is the significant thing, not the thing in which it is coded."

Poor Boulding. He has inadvertently invited a repeat performance. "Describing our interaction that way really gets at what I was talking about," I say. "I mean, you could describe our interaction that way without reference to any conscious states, and as far as the theory of evolution goes, that would be perfectly adequate. We could describe everything I need to do to get my genes transmitted to the next generation without referring to consciousness."

"That's right, yes."

"That's why it seems such a miracle that consciousness exists."

"Yes it is, yes. It's another order of reality."

Glad we got that settled.

Now Boulding resumes the tangent he was pursuing when I interrupted, the upshot of which is that the evolution of human beings can be described as a progression from "know-how" to "know-what."

The classic example of know-how is DNA, which "knows-how" to construct an organism. In the beginning, indeed, DNA was the *only* kind of know-how. But after a time, one of the things that some kinds of DNA knew how to do was construct a large and complex brain, large and complex enough to create, store, and transmit its own know-how. Thus, the brains of some lower primates knew how to take a long stick and clobber prey, and they could ship this knowledge down through the generations, from brain to brain to brain.

As cultural evolution gained momentum, know-what was born. Know-what is descriptive knowledge, and it entails conscious understanding, like the understanding—attained, presumably, somewhere between a chimpanzee's degree of complexity and a human's degree—that prey will drop dead if clobbered. "Once we get know-what," Boulding says, "this produces a profound change in the evolutionary process. With the development of the human race, evolution goes into what I would call a gear change."

Evolution's passage from know-how to know-what could also be described as a progression toward more explicit descriptions of reality. Know-how, after all, is *implicitly* descriptive; DNA indirectly reflects the properties of its environment by virtue of its exploitation of them. Crab grass DNA, in orchestrating photosynthesis, is implicitly noting that energy can be stored by using a particle of sunlight to cleave an H_2O molecule—and thus is implicitly describing the affinity of hydrogen atoms for oxygen atoms. The DNA is also, for that matter, implicitly noting that sun shines and water flows on this planet.

This is a bit metaphorical, of course—a bit like saying that a smooth rock on a creek bed describes the relentless flow of water over it, or that the trees above Baker Beach, at the base of the Golden Gate Bridge—leaning, as they do, away from the water—describe the wind's prevailing direction. But let's face it: in a sense these things are true. In a sense, each person's DNA is a record of the last couple of billion years of the earth's history. Theodosius Dobzhansky probably did not mean to be taken metaphorically when he wrote that "natural selection is a process conveying 'information' about the state of the environment to the genotypes of its inhabitants."

With human beings, and their know-what, descriptions of the environment become explicit. We *understand* the affinity between hydrogen and oxygen. (In fact, we understand, in broad outline, how this affinity came to be implicitly recorded in the DNA of crab grass.) And our understanding of molecules and atoms and subatomic particles—our know-what—is itself stored in molecules and atoms and subatomic particles, distinctive physical configurations in our books and our brains. It is easy, in the light of this progression from know-how to know-what, to appreciate an observation of Teilhard de Chardin's: "The history of the living world can be summarised as the elaboration of ever more perfect eyes within a cosmos in which there is always something more to be seen."

What amazes Boulding is that the universe's awareness of itself grows out of randomness. Random genetic mutations led one-celled organisms to process meaningful information, and random mutations then led these cells to share information so intimately among themselves as to constitute multicellular organisms. Random mutations led the descendants of these organisms, such as ants and prairie dogs and primates, to share information also, and thus to carry the processing of information to the social level. One species of primate even invented elaborate artificial information processing and transmitting systems, which are further integrating the most impressive system of information processing ever to appear on this planet—and the only one ever to encompass it.

Boulding, having reached the end of his tangent, sums it up. Evolution, he says, is "an enormously stochastic process that clearly has a prejudice—towards complexity and consciousness and all that." He pauses. "It's *almost* enough to make a case for creationism." He laughs ambiguously.

This line does not win my hearty assent. I am all for feasible reconciliations of religion and science, but I'm not especially enthusiastic about creationism, inasmuch as it contradicts the most plainly beautiful creation of scientific thought yet, the theory of natural selection. So I ask him about a less radical, less destructive reconciliation of science with some notion of divinity. In a way, I say, I find an intelligence that could create the human species out of thin air less awesome than an intelligence that could create a universe that would give birth to a process as subtly powerful as natural selection—a process that, given a few self-replicating molecules and a few billion years, leads to a theory of itself. (I will later find out how far I am from being the first to make such a point. Shortly after receiving a complimentary copy of *Origin of the Species*, Charles Kingsley, an Anglican clergyman and a naturalist, wrote, in a letter to Charles Darwin, "I have gradually learnt to see that it is just as noble a conception of Deity, to believe that He created primal forms capable of self-development into all forms needful *pro tempore* and *pro loco*, as to believe that He required a fresh act of intervention to supply the *lacunas* which he himself had made. I question whether the former be not the loftier thought.")

Boulding agrees—with me and, by implication, with Kingsley—and takes pains to erase any misinterpretation of his remark about creationism. "Well, these are metaphors, aren't they? That is, the creation metaphor is attractive, because we are ourselves creators. I mean, we make pots. We create all these artifacts. But that metaphor is quite inappropriate for the evolutionary." He pauses. "Well, how should I put it? I'm quite convinced that there is creation, but creation doesn't necessarily imply the metaphor of a creator in the sense that we make a pot, you see. The creator is much more complex and much more subtle a process. The actual process is beyond our comprehension."

Boulding is now more vibrant than I've ever seen him. His still-young blue eyes, set off sharply by his white hair and the weathered skin around them, are intense with interest, and they infuse his words with energy. I can see how anyone who believes in prophets would find him an attractive candidate.

What he's saying sounds fine to me. Personally, I would be willing to settle for the idea that God is in some sense immanent in everything around us, and is not, in fact, a guy in a long gray beard and white robes who keeps score. That is at least as reassuring as agnosticism, and less depressing than atheism. Still, I don't think it will make many people very happy. People seem to want a God who gives meaning to their lives straightforwardly, a God who imparts value by evaluating. They want some of their behavior to be good and some of it bad. They want rewards and punishment, heaven and hell. They want to play for high stakes. And they want to know what stakes they're playing for; unlike Boulding, they equate faith with certainty. I ask Boulding if he can say for *certain* whether there's anything, you know, out there.

"It's a good question, and I don't quite know the answer to it. Actually, I'd put it this way. What I feel very certain about is the existence of potential, you see." He recalls Buckminster Fuller's remark that "God is a verb," and then heads off into the Great Beyond, saying the kinds of self-evident yet cryptic things that people say when trying to talk about the unknowable. "There is this enormous potential, obviously, otherwise we wouldn't be here at all. What leads to the realization of potential is a profound mystery. . . . But, uh, I think perhaps the real core of the religious experience is the experience of potential." He pauses. "Well, the potential is much larger than what is realized, let's put it that way." He pauses again. "On the other hand"—he looks at me with significance—"we can in a sense ally ourselves with the potential."

Ally ourselves with the potential? Somehow I had hoped that when Boulding finally got around to revealing the meaning of life, it would be something a little more elaborate than "Go with the flow."

But if "Go with the flow" it is, then "Go with the flow" it is. The flow, it seems, is evolution's "directionality": the growth of complexity and consciousness, the periodic integration of parts into wholes. Going with it means, at this particular point in organic history, a little more "gentling": less hatred; fewer wars; the subordination of selfishness to interpersonal, interracial, international harmony; the diffusion of the ego in the form of love. It is noteworthy—and Boulding notes it—that all religions stand in some sense for an identification with the whole. (The etymology of *religion* is cloudy, but apparently the word came, like the word *ligament*, from the Latin *ligare*: "to bind together.") Christianity, Judaism, and Islam, as well as the religions of the East, counsel stringent control of, if not outright dissolution of, the ego, and the extension of fraternal bonds beyond the family. Indeed, religion has been credited—by, among others, Dobzhansky and the psychologist Donald Campbell—with helping to carry social cohesion from the level of the kin-based troop to the levels of the tribe and the nation-state. Of course, in the process, religions have also engendered sectarian squabbling and ignited wars. Still, the essence of the religious creeds is not division but unity. "In the major world religions, the divine is in some sense the totality," Boulding says.

An identification with the whole world, he continues, is furthered, as Teilhard de Chardin anticipated, by modern communications technologies. The image of the fragile blue-and-white sphere we live on, beamed around the world from outer space, brought people of different backgrounds a bit closer together. "I think it is true," he says, "that there's an increasing sense of the world as a totality."

Being Kenneth Boulding, though, he must qualify this generalization copiously: don't forget the local and national cultural idiosyncrasies, the opposing economic philosophies that divide the world in two, the gross economic disparity between the northern and southern hemispheres. And being Kenneth Boulding, he must be ambivalent about what unity there is. He is glad that people of different nations have common cultural ground, but he is sorry that airports everywhere look the same, and that pizza and hamburger joints populate the streets of Tokyo. He has fretted in print about the adulteration of "pure cultures," and he realizes that their preservation is difficult, if not impossible, in the information age. "I do think it's very important to have a level of variety, that this is the only thing that really can give us unity. I've often said, I think of the Catholic Church rather as I do of the blue whale. I'm not a Catholic, and I'm not a blue whale, but I'd feel diminished if either of them became extinct."

Boulding is ambivalent about political unity as well as cultural unity. On the one hand, he believes the nation-state, in its traditionally autonomous and often aggressive form, is rapidly becoming obsolete. A world poised on the brink of various sorts of apocalypse can't afford much national egotism. On the other hand, he worries about the worldwide oppression that might be the price for global harmony of an organic sort. "A world state could be a nightmare," he says. "World tyranny of the worst kind. I've sometimes said I'd rather have refugees than have no place of refuge, you see." Then, sustaining a classically Bouldingesque train of thought, he adds: "There's something to be said for this space colony stuff."

Boulding's ideal world would be in some ways like an organism and in some ways more like an ecosystem, an international ecosystem that could reconcile stability with autonomy by striking a healthy balance among lots of social organisms—governments, corporations, unions, churches. A fairly loose global federalism is what Boulding prescribes: supranational bodies more powerful than the United Nations, but not all-powerful.

Of course, to supplement worldwide political bonds, we'll need to expand and strengthen the integrative system. But, Boulding stresses, that doesn't mean we have to carry things to Teilhardian extremes; he isn't saying we must fervently love people we've never met. He just wants us to treat them the way we would treat fellow members of the Elks Club, or even fellow Americans. It's all right to want to beat the Russians in the Olympics; we just shouldn't want to kill them. Boulding wants to "turn enemies into opponents." His ideal—in many realms and at many levels of organization—is "a perpetual state of unresolved conflict."

All of this helps explain his fondness for the word *potential*. He doesn't like the idea that, as Teilhard de Chardin sometimes seemed to imply, we are headed inevitably toward a rigidly unified world. In fact, he doesn't like the idea that we are headed *inevitably* anywhere. "If teleology involves only a single goal, then it's just as deterministic as determinism," he says. Potential, on the other hand, "always involves freedom."

Kenneth Boulding likes his unity as much as the next man, but he is leery of the ultimate synthesis.

The grandfather clock still reads 8:01. Either Boulding has transported me beyond the dimensions of space and time—a possibility that I'm not prepared to rule out—or this is a dead clock. It is dark outside now, and Boulding is growing tired. His eyes still look young, but his voice has become hoarse. There is time, at best, for a little penetrating retrospection. I ask if over the years it has been a struggle to discipline his meandering mind, to keep it on one subject long enough for accomplishment.

"No," he says simply. Then: "I never worried very much about what people thought about me. That wasn't very important or relevant. I don't say I haven't enjoyed having all these honors and all that, but it isn't something that's in any sense fundamental. I've just pursued my interests and just had fun. In my intellectual life I've always pursued the things that excited me. And if somebody else found it exciting too, that was just fine. I've been lucky this way. I

just read this enormous, ponderous German book on envy. I've forgotten the man's name now. This is something I was very little subject
to. It never bothered me whether anybody else did better than I did.
Maybe it was just that I never liked sports." He laughs. "I couldn't
be bothered with winning anything. I've always liked hiking and
mountain scrambling and just things you do for fun, you see. Of
course, I was never any good at it, either. I could never throw a ball
or hit anything in my life. Every race I ran as a little boy, I'm sure
I came in last. But as I say, this whole business of winning never
interested me at all. Being and doing and having fun, and so on, you
see—and just enjoying life."

But doesn't he give any thought to his legacy? "No. No, I've never
worried about it very much." Suppose he were forced to think about
it; suppose he had to choose a legacy. "Well, I think if people look
back at my work and it cheers them up, then I'll feel I've done well."
He laughs.

I can't help but think that there's something more here, that his
career has been driven by something deeper than the pursuit of
intellectual joy and public amusement. Indeed, he has admitted as
much. Once he told me that he has spent his professional life studying
two questions: First, what does it mean to say that things have gone
from bad to better rather than from bad to worse? Second, how do
we get to better? Surely his devotion to human betterment represents
a link between his religious convictions and his work. Surely he will
talk about this if I broach the subject. It is a delicate subject, of
course, one that should be broached with grace. I am up to the
challenge; I ask, "Do you sometimes have the feeling that you're in
academia as a sort of undercover agent of God?"

He laughs—whether with me, at me, or in uneasy self-defense I
can't tell—and doesn't answer immediately. In a too-late attempt to
sand down some of the question's rough edges, I add, "It's so rare to
find a well-known scientist or social scientist who is avowedly religious." He laughs again, and I laugh, and he still says nothing. I
persist: "Have you ever looked at it like that?" There is no response
for several seconds.

"No, I don't think so," he finally says. "No."

But he looks away as he answers, and after a few more seconds he
notes that this is a very personal question I've just asked.

We talk for a minute or so more. I press him for a few final lines of spacy speculation about evolution, information, and the meaning of life. "Well, as I say, the development of potentiality has some directionality about it, to put it in secular language," he says conclusively. "Well, I think I'm pretty exhausted. I think I'll retire."

While steering me toward the door, he remembers that he had promised to give me a Quaker pamphlet about international relations. We walk up the stairs, and he opens a door at the end of the hall. An aged double bed, with red-checked spread, covers more than half the floor. Even a room this small is only sparsely populated by Boulding's possessions: a few hanging clothes, a Walkman containing a Benny Goodman cassette, a book called *The Economics of Time and Ignorance*—which he did not write, though he probably would have, given enough time. An image of uncertain identity—drawn by his two-and-a-half-year-old granddaughter—graces the closet door. Boulding says something, self-consciously, about how small the room is. But I doubt it really cramps his style, and I'm sure he could get by on less.

He hands me the pamphlet. "Mending the World," reads the title page. There is a sketched map of Alaska and Siberia, bound to one another by a piece of giant thread that has been pulled eastward across the Bering Strait and through Alaska by a giant needle. Below the sketch is the subtitle, "Quaker Insights on the Social Order," and the name of the author: Kenneth E. Boulding.

When Boulding leaves New York in a few weeks, the book he had planned to write at the Russell Sage Foundation—*The Logic of Love*—will still be unwritten. Its bits and pieces will still sit in a folder, and when he pulls the folder from a desk drawer, slips of paper will still fall out of the ends. The integrative system continues to defy integration, and I suspect that it will do so for the rest of his life. What's more, I suspect that he suspects that, too, and would freely acknowledge the possibility if I were tactless enough to raise it. And I suspect that the possibility doesn't trouble him a whole lot, and that tonight, on this sturdy old bed, he will sleep peacefully.

EPILOGUE

THE MEANINGS OF
THE MEANING OF LIFE

Lots of extremes have been gone to in search of the meaning of life. Mountains have been climbed; worldly pleasures have been eschewed; meaningless phrases have been chanted for hours in dark rooms; psychoactive chemicals have been ingested; Frisbees have been thrown. Far be it from me to question the value of any of these approaches. But, for the benefit of those who seek moderation in all things, I would like to mention an alternative. If you want to know about the meaning of life, just do what you do when you want to know about anything else: look it up. Specifically, get out the *Encyclopedia of Philosophy* and turn to the article on page 467 of volume four: "Life, Meaning and Value of." There—under the subheading "The meanings of 'the meaning of life' "—it is written that the phrase "meaning of life" traditionally has been used in two senses, both closely connected with the concept of purpose.

First, the meaning of life has been tied to *divinely imparted* purpose. As the encyclopedia puts it, "Sometimes when a person asks whether life has any meaning, what he wants to know is whether there is a superhuman intelligence that fashioned human beings along with other objects in the world to serve some end—whether their role is perhaps analogous to the part of an instrument (or its player) in a symphony." The article then goes on to suggest that modern science sheds little light on this question, and cites an observation by the astronomer Fred Hoyle—that we find ourselves in a "dreadful situation," with "scarcely a clue as to whether our existence has any real significance."

For the sake of balance, it should be noted that there are people—smart people, even—who think science *does* offer evidence of significance. They believe that the sheer elegance of natural selection, the logical beauty that enraptured the young E. O. Wilson, implies superhuman design; that only a vast intelligence could invent a process of such power and simplicity that it creates beings able to understand the way it works but at a loss to explain exactly how *they* work.

There is no denying the empirical core of this claim—the potent grace with which natural selection has translated the occasionally erroneous replication of molecular information into ever larger examples of order and, moreover, complexity. At various levels of organization, evolution's invisible hand has forged an ironic equation between unyielding self-interest—whether of genes, cells, or multi-celled organisms—and collective harmony.

One sign of the elegance of evolution is that, in retrospect, it all appears more or less inevitable. The impulsion of organic coherence upward through level after level of organization seems almost to follow necessarily from a few basic principles. The first principle, needless to say, is natural selection. Another is kin selection—which, actually, is a corollary of natural selection, and therefore doesn't really count. Then there are principles of economical operation, such as division of labor. And at rock bottom is a principle more basic than any of these, a physical law that was hard at work long before organic laws existed: the second law of thermodynamics. It is the second law that, early on, punished with extinction strands of DNA that failed to surround themselves with walls against entropy, and it is the second law that insisted that these flexible fortresses process information. As if that weren't enough, this information, under the metaphysical laws governing this universe, seems to bring conscious experience when processed with sophistication (and thus gives life a *certain* meaning, whether or not it is a kind recognized by the *Encyclopedia of Philosophy*). In short, a few basic principles that prevail in this universe seem to imply almost inexorably the construction of lots of machinery, some of which is large and complicated and capable of having a good time.

This thumbnail summary of the logic behind life deserves to be greeted with skepticism; things often appear inevitable in retrospect, but there's usually no way of finding out if they were. And besides—to return to the origin of this tangent—even if we grant that the creation and evolution of life has been an extremely elegant process, elegance does not unambiguously imply design. To be sure, in *our* world elegant things typically are the handiwork of intelligent things. But, as Ed Fredkin noted, our world may be nothing like any larger worlds encompassing it. So if life's having some meaning is dependent on its having an intelligent creator, we're out of luck; science, by itself, cannot support any position on this question more consoling than agnosticism. Fred Hoyle, sad to say, got the story essentially right.

Purpose seekers may be cheered, then, to hear about the second sense in which, according to the *Encyclopedia of Philosophy*, life can have meaning: whether or not life *in general* has meaning, individual lives are said to have it if they are guided by commitment to an overarching cause. A paraplegic who wheels himself across the country to dramatize the plight and potential of the handicapped, a missionary who lives at the subsistence level in New Guinea, a politician who tries tirelessly to spread creeping socialism—these people's lives could be said to have purpose, and therefore meaning, in one traditional sense of the term.

One problem with this equation between meaning and commitment to a cause is that, taken literally, it attributes meaning to lives that almost everyone considers meaningless. Hedonism, for example, is a cause of sorts; Sid Vicious, of the Sex Pistols, demonstrated truly steadfast devotion to sensual pleasure right up until the drug-assisted deaths of his lovely girlfriend and himself. But somehow I don't consider his life-style a paragon for youths in search of meaning. And I have similar reservations about many ostensibly more upstanding people—more circumspectly hedonistic people, people whose lives are devoted with great self-discipline entirely to the acquisition of, and indulgence in, material wealth.

Clearly, then, we're looking at a dilemma: on the one hand, we can't say with any confidence that life on earth has *divinely imparted* purpose, and therefore meaning; on the other hand, it just doesn't make sense to deem meaningful any life with *any* sort of purpose.

Some people get around this problem with a compromise. Okay, they say, we can't be absolutely, positively sure that there's any intelligence behind evolution; but, just for the hell of it, let's sort of assume that the basic direction of evolution is worth sustaining, and that helping to sustain it is therefore a purpose that can bring meaning to a life. Ed Fredkin, for example, who thinks it pretty likely that *something* intelligent is ultimately behind evolution, but isn't sure what, said, when asked about the meaning of life, that our purpose— *mission* is the word he used—is to create artificial intelligence; his reasoning was simply that this is the next logical step in evolution. Even E. O. Wilson accepts a kind of derivation of individual purpose from collective direction: human cultural evolution has carried us away from religion and toward science, so, he says, we might as well accept our fate and get enthralled by science.

Kenneth Boulding also thinks we should go with the flow, but he sees a Teilhardian flow, an integration of parts into a greater whole, the subordination of human and national egos to global harmony. According to this view, we can bring meaning to our lives by sponsoring orphans in India, or by seeking pen pals in China, or by traveling to the Soviet Union as civilian ambassadors of goodwill, or by housing students visiting from Africa, or by teaching our children that all human beings are human, and only human—or, really, just by being nice to people and hoping that niceness is catching.

This Bouldingesque conception of meaning in life typically encounters two objections. The first is applicable to all moral extrapolations from the direction of evolution: it is notoriously fallacious to infer ought from is; the fact that evolution has moved in a certain direction does not imply that it *should* move in that direction, or that we should lend it our support. If we don't know whether there's any intelligence behind evolution, then we can't assume there's any sense in it. (And, besides, we have no way of knowing whether any such intelligence would be benign or malicious.)

The second objection to Bouldingesque meaning is less philosophical and more practical: even if evolution has repeatedly headed toward integration in the past, it is unlikely to continue doing so, because the logic that got it this far has ceased to apply. As Boulding himself has noted, integration is often fostered by external threat. Prairie dog colonies are more cohesive than they would be had they never faced predators; corporations are better organized than they would be in the absence of competitors; and the United States might descend into dissension were it not for the perception of various external threats, notably the Communist Bloc. (A president's Gallup approval rating rises in the midst of international crisis as reflexively as a cellular slime mold coalesces in the face of drought.) Skeptics of Bouldingesque meaning say that, barring well-armed visitors from another solar system, the world will never face an external threat sufficient to induce its cohesion.

But those who hold out faith for global harmony have an interesting reply to the skeptics: as the cellular slime mold demonstrates, living, breathing threats are not the only kind that foster cooperation; all that matters is that the threat be mutual. And the modern world is rife with threats common to many, if not all, nations: impending nuclear holocaust, ozone depletion, acid rain, air piracy and other terrorism, contamination of tuna, nuclear meltdowns, the economically unhealthy diversion of resources to armament. These threats, though not of an origin external to the planet, can, nonetheless, best be battled through concerted international action. They may not arouse a visceral global patriotism, but they add power to the logic of cooperation as surely as would invaders from Mars.

There is evidence—scanty evidence, for the time being, but evidence nonetheless—that this power is having effect. Though channels of communication do not cross the Iron Curtain as profusely as we might like, what channels there are have resulted largely from grave global threats: nuclear war, nuclear meltdown, air and water pollution. Even terrorism has brought passing expressions of sympathy across the curtain and tentative steps toward preventive cooperation. To the extent that dialogue bears fruit in these areas, it will move the planet, albeit incrementally, toward the subordination of national behavior to supranational agreements, and hence toward a modest degree of superorganic unity.

The reason it will do so—if in fact it does—is because these global threats preserve the equation that got us here in the first place: the invisible hand's translation of individual interests into larger order. Just as genetic self-interest manifested itself in the form of organismic and then social coherence, and human self-interest spawned organizations of much larger scope, including nations, various common threats show signs of translating individual and national self-interest into global order.

This equation of self-interest with global harmony happens to quite handily demolish not only the practical objection to Bouldingesque meaning but the philosophical one as well—the fallacy of inferring ought from is. For if it is in our interest to sustain the direction of evolution, then we needn't infer this purpose blindly from that direction; we can pursue it just because it makes sense.

So the evolutionary "logic" that has driven life to violate the spirit of the second law all along is now suddenly literal: not only is self-interest equated with a larger order; we are *thinking*—consciously, explicitly, and sometimes even rationally—about this equation and trying to realize it. Of course, history is full of evidence that, in human affairs, logic does not always prevail. We may yet fall prey to the dark vestiges of our evolutionary past and blow the whole planet up or do something equally stupid. This capacity to screw things up is something that Boulding would no doubt stress. And he might add that it is what imparts meaning to our actions ultimately.

Now for the bonus question: What does it mean that some fairly reasonable (as these things go) attempts to extract purpose and meaning from evolution bear results remarkably like longstanding doctrine of the world's great religions? Is it just a coincidence? Fredkin probably would say so. Wilson probably would say that it has something to do with the pragmatic criteria of selection in the genetic and cultural evolution that gave birth to religion. Boulding probably would smile enigmatically and say something vague and suggestive. Personally, I don't know what to think. But I think about it often.

SELECTED BIBLIOGRAPHY

Beniger, James R. *The Control Revolution: Technological and Economic Origins of the Information Society.* Cambridge, Mass.: Harvard University Press, 1986. Before tracing the roots of the information age back to the industrial revolution, Beniger sets this subject in its most fundamental context by discussing the second law of thermodynamics, its relation to life, and the role of information in organisms and societies. This boundless range is reminiscent of general system theory, though Beniger doesn't use that label. Table 3.2 is especially worth contemplating.

Bennett, Charles H. "Demons, Engines and the Second Law." *Scientific American* 257 (1987), no. 5. What Ed Fredkin's (and Bennett's) reversible computers tell us about why Maxwell's Demon can't win.

Bennett, Charles H., and Rolf Landauer. "The Fundamental Physical Limits of Computation." *Scientific American* 253 (1985), no. 1. This discussion of reversible computing—including Fredkin gates and the billiard ball computer—is not exactly beach reading, but it's less dense than the technical literature on the subject.

Bertalanffy, Ludwig von. *General System Theory.* New York: George Braziller, 1968. A mostly accessible and not overly long introduction to the field by its founder.

Bonner, John Tyler. *The Evolution of Culture in Animals.* Princeton, N.J.: Princeton University Press, 1980. A nicely written (and illustrated) book that shows how far we are from being the only cultural beings.

Boulding, Kenneth. *The Organizational Revolution: A Study in the Ethics of Economic Organization.* New York: Harper and Brothers, 1953. One of Boulding's better focused efforts, but not at the expense of his eclecticism and literary flourish.

———. *Ecodynamics: A New Theory of Societal Evolution.* Beverly Hills: Sage Publications, 1978. Boulding considers this a grand summing up. The word *theory* in the subtitle suggests something less diffuse than the book turns out to be, but on a per-insight basis this would be a bargain at twice the price.

Burnham, David. *The Rise of the Computer State.* New York: Random House, 1983. Burnham is a bit hypersensitive in his concern about big brother's emergence in the United States, but the danger is real enough so that it's good to have somebody like him sounding alarms.

Cairns-Smith, A. G. "The First Organisms." *Scientific American* 252 (1985), no. 6. How did mere form become a form of life? Cairns-Smith has a very speculative but instructive theory about clay crystals' having been the original information copiers, precursors of DNA.

Campbell, Jeremy. *Grammatical Man: Information, Entropy, Language and Life*. New York: Simon and Schuster, 1982. Campbell's book is where I first saw explicit reference to the idea of information as a unifying, vertically integrating theme in the sciences and social sciences. More specifically, he puts much emphasis on the explanatory power of information *theory* at different levels of organic organization—an emphasis that many observers have come over the past few decades to consider unwarranted. Whatever the merits of Campbell's view, this is a worthwhile, very readable book.

Crapo, Lawrence. *Hormones: The Messengers of Life*. New York: W. H. Freeman and Co., 1985. An authoritative but brief and accessible introduction to hormones, spiced with some great moments in scientific history.

Dawkins, Richard. *The Selfish Gene*. New York: Oxford University Press, 1976. I don't know of any stronger combination of analytical and expository skill than Dawkins's. This small book conveys better than Wilson's *Sociobiology* the power of the theory of natural selection in explaining animal behavior, including our own. The last chapter is an engaging, idiosyncratic discussion of cultural evolution.

Dawkins, Richard, and John R. Krebs. "Animal Signals: Information or Manipulation?" In *Behavioral Ecology: An Evolutionary Approach*, edited by John R. Krebs and N. B. Davies. Sunderland, Mass.: Sinauer, 1984. Did communication evolve as a way of sharing information or as a means of control? The authors stress the latter scenario (though the two are hardly incompatible).

Dobzhansky, Theodosius. *Mankind Evolving: The Evolution of the Human Species*. New Haven, Conn.: Yale University Press, 1962. After all these years, this book remains on the frontiers of judiciousness. Its topics range from social Darwinism to the relationship between cultural and genetic evolution.

————. "Teilhard de Chardin and the Orientation of Evolution." *Zygon* 3 (1968), no. 3. Dobzhansky here argued that, although Teilhard did not grasp many details of the mechanics of evolution, he understood the essence of natural selection and reconciled his philosophy with it.

Fredkin, Edward, and Tomasso Toffoli. "Conservative Logic." *International Journal of Theoretical Physics* 21 (1982), nos. 3/4. This discussion of reversible computation and its correspondence to reversible physical processes is not for a lay audience, but it merits mention here anyway, for it has the distinction of being a published paper by Edward Fredkin.

Goody, Jack. *The Logic of Writing and the Organization of Society*. New York: Cambridge University Press, 1986. By examining cases drawn from history and from literate and nonliterate contemporary societies, Goody sheds light on, among other things, the impact of writing on the economic, governmental, and legal structures of a society.

Harold, Franklin M. *The Vital Force: A Study of Bioenergetics*. New York: W. H. Freeman, 1986. In the preface, Harold writes, "What breathes life into the jumble of complex molecules is expressed in deceptively familiar words: *energy*, *work*, *information*, *order*." The rest of the book is about how exactly this life is breathed in, an inquiry that joins thermodynamics and biology. I only read the first and the last two chapters—"Energy, Work, and Order," "Signals for Communication and Control," and "Morphogenesis and Biological Order"—but they were worth the price of admission. This is essentially a textbook, but a fairly readable one.

Hofstadter, Douglas R. "On Viral Sentences and Self-Replicating Structures." *Scientific American* 248 (1983), no. 1. Reprinted in *Metamagical Themas*, by Douglas R. Hofstadter. New York: Basic Books, 1985. Building on Dawkins's idea of memes, Hofstadter brings rigor to the idea of viewing written sequences of information as living things that, like genes, can succeed to various degrees (including zero) in replicating themselves. Hofstadter's well-known expository skills are put to good use here.

Hofstadter, Richard. *Social Darwinism in American Thought.* 1944. Reprint. Boston: Beacon Press, 1955. The chapter on Herbert Spencer valuably demonstrates how careful we have to be with metaphors such as "superorganism" and phrases such as "survival of the fittest," as they are subject to sordid political exploitation. It also demonstrates, I think, that some of Spencer's more worthwhile insights—particularly regarding the relationship between evolution and religious doctrine—were lost in the outcry that his ethical philosophy and political orientation eventually provoked.

Jacob, François. *The Logic of Life: A History of Heredity.* New York: Vintage Books, 1976. The last chapter, "The Integron," should be required reading in every introductory biology course—and, for that matter, in every introductory sociology course.

Kerman, Cynthia Earl. *Creative Tension: The Life and Thought of Kenneth Boulding.* Ann Arbor: University of Michigan Press, 1974. This exhaustively researched, well-written, and insightful book was of immense help to me, and would be a good read for anyone interested in Boulding. Its unusual organization is apparently due to the fact that it began as a Ph.D. dissertation, with the aim of dissecting Boulding's mind and character rather than documenting his life chronologically.

Machlup, Fritz. *The Production and Distribution of Knowledge in the United States.* Princeton, N.J.: Princeton University Press, 1962. A decade before Daniel Bell wrote *The Coming of Post-Industrial Society,* Machlup identified an economic sector roughly synonymous with what is now called the information sector and noted that it was growing by leaps and bounds.

Machlup, Fritz, and Una Mansfield, eds. *The Study of Information: Interdisciplinary Messages.* New York: John Wiley and Sons, 1983. More than forty contributors cover the waterfront; there are whole sections on cybernetics, information theory, general system theory, linguistics, artificial intelligence—even library science. Much of the writing is very dense, but the book's sheer scope makes it a unique resource.

MacKay, Donald M. *Information, Mechanism, and Meaning.* Cambridge, Mass.: MIT Press, 1969. Reading this series of post–World War II papers is a good way to get a feel for the evolution of scientific thinking about information. MacKay's ideas about meaning would probably earn Charles Peirce's seal of approval, and they are set out lucidly, though in other places the book is less accessible.

Marler, Peter, and Richard Tenaza. "Signaling Behavior of Apes with Special Reference to Vocalization." In *How Animals Communicate*, edited by Thomas A. Sebeok. Bloomington: Indiana University Press, 1977. An acoustical and functional analysis of the dozen or so utterances that typically make up the vocabulary of apes, with special attention to chimpanzees, our nearest living relatives.

Mayr, Otto. "The Origins of Feedback Control." *Scientific American* 223 (1970), no. 4. Machines, it turns out, have been processing meaningful information for a long time—even before the first century A.D., when Hero of Alexandria invented an automated wine dispenser.

Miller, James Grier. *Living Systems*. New York: McGraw-Hill, 1978. General system theory at its most ambitious and exhaustive. Chapter 6, on page 203, is "The Cell." Chapter 12, on page 903, is "The Supranational System." In between is the rest.

Monod, Jacques. *Chance and Necessity: An Essay on the Natural Philosophy of Modern Biology*. New York: Alfred A. Knopf, 1971. Monod sets out to explain purposeful behavior without reference to spooky, immaterial forces, and in the process takes Henri Bergson, Teilhard de Chardin, and others of their ilk to task for sloppy thinking. Monod, a Nobel Prize–winning biochemist, touches on Maxwell's Demon and notes that organic coherence in the face of the second law of thermodynamics rests ultimately on the ability of protein molecules to " 'recognize' other molecules," thus exercising a "discriminative (if not 'cognitive') function." Unfortunately, the book presupposes an understanding of Monod's specialized terminology, so the reader sometimes has the feeling of having walked in on a fascinating seminar thirty minutes late.

Peirce, Charles S. "How to Make Our Ideas Clear." In *Philosophical Writings of Peirce*, edited by Justus Buchler. New York: Dover Publications, 1955. This is the article (first published in *Popular Science Monthly* in 1878) in which Peirce introduced his pragmatic conception of meaning.

Roth, Jessie, and Derek Le Roith. "Chemical Cross Talk." *The Sciences* 27 (1987), no. 3. The authors' "unifying theory of intercellular communication" explains why so many plant derivatives, from morphine to ephedrine, systematically affect the functioning of animals.

Rubin, Michael Rogers, and Mary Taylor Huber. *The Knowledge Industry in the United States, 1960–1980*. Princeton, N.J.: Princeton University Press, 1986. A meticulous updating of the many statistics found in Machlup's 1962 book, compiled after his death. It consists mainly of tables (and exegeses on them) that together depict in great detail the evolution of the information society since 1958.

Schmandt-Besserat, Denise. "Oneness, Twoness, Threeness." *The Sciences* 27 (1987) no. 4. This account of how the ancients invented numerals (and letters) highlights the correlation between a society's complexity and the complexity of its symbols.

Schrödinger, Erwin. *What Is Life? and Mind and Matter*. New York: Cambridge University Press, 1980. The first of these two long essays, originally published in 1948, was among the first accessible writing about the relationship between the second law of thermodynamics and life. Incidentally, the second essay (first published in 1958), with its fairly weird speculation about consciousness, foreshadowed the physics-and-mysticism boom of the 1970s and 1980s.

Sebeok, Thomas A., ed. *How Animals Communicate*. Bloomington: Indiana University Press, 1977. Thirty-eight articles, covering more than a thousand pages, on everything from "The Phylogeny of Language" to "Communication in Sirenians, Sea Otters, and Pinnipeds."

Shannon, Claude E., and Warren Weaver. *The Mathematical Theory of Communication*. 1949. Reprint. Chicago: University of Illinois Press, 1963. Shannon's famous paper, preceded by Weaver's summary introduction. The former is for math whizzes, the latter for the rest of us.

Simon, Herbert. "The Architecture of Complexity." In *The Sciences of the Artificial*, by Herbert Simon. Cambridge, Mass.: MIT Press, 1969. Why does almost everything complex, whether plant, animal, or artifact, consist of "subassemblies"—distinct components comprising still smaller distinct components? Simon has a plausible answer.

Stent, Gunther. "Explicit and Implicit Semantic Content of the Genetic Information." In *The Centrality of Science and Absolute Values*. Proceedings of the Fourth International Conference on the Unity of the Sciences. New York, 1975. Stent discusses a somewhat metaphorical sense—which I mentioned briefly in chapter 10 and then ignored in favor of a more pragmatic sense—in which DNA can be said to have "meaning." This paper inspired Douglas R. Hofstadter's better-known treatment of the same topic—which is also worth taking a look at (in *Gödel, Escher, Bach: An Eternal Golden Braid*).

Teilhard de Chardin, Pierre. *The Future of Man*. New York: Harper Torchbook, 1969. If you have time for only one book by Teilhard, leaf through the essays in this one (first published in English in 1964; in French in 1959) rather than tackle his better known but more obscure *The Phenomenon of Man*.

Toffoli, Tomasso, and Norman Margolus. *Cellular Automata Machines: A New Environment for Modeling*. Cambridge, Mass.: MIT Press, 1987. These two protégés of Fredkin's discuss cellular automata generally, the newly available CAM-6 cellular automata machine in particular, and the use of cellular automata to model parts of this universe and to create imaginary universes.

Tomkins, Gordon M. "The Metabolic Code." *Science*. 5 September 1975. As the article's subtitle puts it, "Biological symbolism and the origin of intercellular communication is discussed."

Tribus, Myron, and Edward C. McIrvine. "Energy and Information." *Scientific American* 225 (1971), no. 3. Though largely about thermodynamics, this article doesn't shy away from fundamental truths applicable at the higher levels of organization; for example, "It takes energy to obtain knowledge and . . . it takes information to harness energy."

Wheeler, William Morton. "The Ant Colony as Organism." *Journal of Morphology* 22 (1911), no. 2. A point well taken.

Wiener, Norbert. *Cybernetics: or Control and Communication in the Animal and the Machine*. Cambridge, Mass.: MIT Press, 1982. In this paperback edition, the original 1948 text is supplemented with two chapters on learning and brain waves, added in 1961. Most of the book is tough sledding, but the more intelligible twenty-nine-page introduction captures the spirit of cybernetics.

————. *The Human Use of Human Beings*. New York: Avon Books, 1967. Cybernetics made simple, along with Wiener's then-growing concerns about the political misuse of technology.

Wiener, Norbert (with Arturo Rosenblueth and Julian Bigelow). "Behavior, Purpose and Teleology." *Philosophy of Science* 10 (1943), no. 1. "All purposeful behavior may be considered to require negative feed-back"—and other ambitious insights.

Wilson, Edward O. *On Human Nature*. Cambridge, Mass.: Harvard University Press, 1978. Short (by Wilson's standards) and very sweet. Consistently readable and interesting, but idiosyncratic in substance, and thus not a proportional representation of the light shed by Darwinism on human behavior.

————. *Sociobiology*. Cambridge, Mass.: Harvard University Press, 1975. Even if he had never mentioned human beings, this would be important as a reference book on animal behavior. Also available in an abridged, paperback version.

INDEX